An Eye Opener to the Impartial Eye

Activate your Immortal Intelligence

Discover who is the Singularity
the Source behind the Universe?

All You Need to Know!

Edited by the author

Mohamed-Rashid Abdullahi Osoble

Contents

Dedication i

Acknowledgements iii

Introduction 1

What is Intelligence? 8

Dual Mechanism of Intelligence 10

    Mortal Intelligence 14

    Immortal Intelligence 34

The Immortal Intelligence Engages with Selfanism 48

The Immortal Intelligence Engages with Atheism 53

The Immortal Intelligence Engages with the Big Bang Theory 72

The Immortal Intelligence Engages with Neutralism. 91

The Immortal Intelligence Engages with Agnosticism 96

The Immortal Intelligence Recognizes 'Impactism' 98

The Immortal Intelligence Plays the Shell Game 108

The Controversy of One Singular God Why Many Religions 113

The Ultimate Meaning of Freedom 131

The Immortal Intelligence Requires Primary Evidence 'Data' 148

The Immortal Intelligence Engages with Hinduism 151

The Immortal Intelligence Engages with Buddhism 155

The Immortal Intelligence Engages with the Hebrew Bible, Moses and Judaism 159

The Immortal Intelligence Engages with The New Testament, Christianity And Jesus Christ 188

Primary Evidence: Science versus Bible 229

    The Age of the Universe 229

    The Flat Earth 232

    The Chronology of Creation 236

    The Heavans are Held by Pillars 238

    Pillars Hold Planet Earth 238

    Drinking Poision is not Hazardous to Followers 238

    Infectious Deseases 240

Women Discrimination in Childbirth 241

Women Discrimination in Infidelity 242

Jesus versus Trinity 244

The Immortal Intelligence Engages with Islam, Muhammad and the Koran 269

Primary Evidence: Science versus Koran 302

Expansion of the Universe 309

Embryology 310

The Beginning of the Universe – the Big Bang Theory 314

Life is based on water 315

Timescale of the Universe: Earth time (Human's time) versus Cosmic Time (God's time) 317

The Sky's role of protective functionality 320

When the Koran Corrected Science and the Bible 321

Iron 'not' an earthly material 323

The Double Entry Preservation of the Koran as a Wondrous Evidence 325

The Beginning 332

Dedication

I thought about the title dedication; it did not pose a challenge and was a no brainer. I said to myself let me dedicate this work to truth seekers, but who is that? Everybody, at some level, is a truth seeker, is it not? Chasing after the dream life, getting a partner, live happily ever after, the normative preset goals of life and the like. While that is fundamental and admirable, the truth is not however, just as life itself, a script we ourselves write and run. If you have any doubt, how many times do you see the unexpected twists and turns of life and times appear before us collectively and individually? Could anybody likely have foreseen the world would come to a standstill 'immobile' in the year 2020 by something the human eye cannot even directly visualize called the COVID-19 virus?

The moment you realize the afore reality is indeed your first initiation 'step' to scratch the surface of what it is that makes the ' truth' before you even start searching it. As simple as the question seems, it is deeper and more complex than that! But not because the truth itself is too complex to be attainable but the way in which it is packaged and proposed to us makes it that way.

Therefore, this book is not a mere dedication to truth seekers. Oh no, my dear friend! It is for those who have the courage to verify the perception of the truth which was packaged and given to them! Question their own reason and observation of the truth, those who have the audacity to search certainty in an uncertain world, unless you are a delegator who transferred his/her intellectual reason on someone else, who ever that may be, in which case you should strip out your title of humanity from yourself and might as well call yourself an artificial intelligence (AI) or pretentious 'fake' person as you should already know this world, your life and mine, are playing out in an uncertain way, while we have somewhat control yet we do not. Nothing, on the whole, would stick to the script!

To sum up, the book you are about to engage with is a profound thought provoker, the level of real mindset one awakens his/her last 24 hours, if tomorrow were your last 24 hours, what would matter? Would it be your career, your wealth, your boss, be it the head of state or CEO of a top 500 Corporation?

A disclaimer: expect as a reader to be challenged, provoked, even enraged at some level while engaging with this book, particularly when the truth gets uncomfortable, including myself, wishing sometimes the requirement for the truth was different or absent altogether, as we all wish every tomorrow to be the beginning of a holiday but it is not realistic. Furthermore, to hear it from stranger like me, 'Mohamed' makes it all the more odd. Is it not the frequent case that it takes an outsider to point out the issues more so than an insider? Further, this might be a test of objectivity on your part. However, rest assured, no claim or statement presented in this book is without evidence, logic or reason. Furthermore, the book addresses a wide range of demographics and belief norms; hence it is recommended to hold your judgments until you reach the end.

Acknowledgements

First and for most, I herewith give acknowledgement to the source behind the universe and, by extension, myself and you, and all that exists. I chose not to name who my research concluded that being is because any fixed answer requires justifying its end result, otherwise, it would amount to plain wishes or blind faith 'indoctrination' as practiced on us since the beginning of time. But I promise, you would be surprised if it is not 'who' it is, then how this is reached, or better still, how it was close to you that all you had to do was to open your 'impartial eyes'. For me, finding an answer to this question is the victory above all victors; you will see what I mean once you read the book!

Secondly, I would like to acknowledge 'Pantheon'; Miss Helen Cooper and my editor, Mr Sebastian Stone, not just for formalities but they were the first audience who appraised my work and reciprocated their early thoughts that above all "it is refreshing to see unfiltered approach in writing". Further, despite, from commercial point of view, I thought to myself to water down some of the rawness in my style, I retained my voice and along with its rawness simply because I learned timidness or shyness, while admirable in general may hinder the full fabric of the authentic truth to surface hence clarity and the bravery to take risk to implement that end was of the essence.

With that said, that is not to say I am for defaming or disrespecting others. As such, I respect and tolerate all people's different cultures and customs. Through my weaknesses, I acknowledge that I am not God. Therefore, who am I to judge others? I absolutely practice judging others is not my place whatsoever, to see why that is, please read the book!

About the Author

The author combines in his book philosophy, religion and science in which he promotes evidence and reason to guide us to any conclusions that we hold. As time goes which usually means more facts-based evidence emerges, scientific or otherwise, our views must change to the facts. Therefore, as such, we are not static in one position at all times. We are always learning and developing individually and collectively as well as cumulatively from each other.

For humanity to maintain its advancement and civilization there must be freedom of ideas to flourish even when it differs from the cultural biases we are accustomed. The benchmark with which humanity as a whole is to be measured is the ability to solve disputes peacefully and maintain overall peace in which we all have a role to play:

"You are not only responsible for a global environmental footprint but more preciously global 'peace' footprint by doing your part, however 'incremental".

Introduction

I have never been a big fan of arbitrary following of traditions for the sake of customs, or the sake of ancestral ethos, or because the society does it in a certain way. As for me, conformity has always been a process and prerequisite of agreement from my intellectual justification or otherwise; it has always been a process which I am 'temporarily' putting up with, a compulsory requirement such as school or duties at work etcetera. If we were blind adherents to tradition, then we would be living in caves like the 'caveman' or earliest 'homo sapien' generations of mankind.

Therefore, if you possess a like mind as mine, henceforth intellectual acceptance is a requirement prior to blindly jumping on the back of the wagon. Then you find yourself to be likely at home in this book. The aim of this piece—whether it becomes a book, essay, or something else—was not to serve as a fairytale or a toy for fictional happiness, nor was it intended to be solely a money-making story, as we are accustomed to seeing from Hollywood.

This content addresses factual realities that demand an organic and raw storytelling approach. It aims to be a truth-depicter and a buried fact-digger rather than a polished commercial tale. Unlike Hollywood's infamous disclaimer, "based on a true story but tailored for commercial purposes, " which tells the audience upfront that the facts have been tempered, this work seeks to present the truth without romanticizing or sugar-coating the facts.

As we will see, this has become 'unfortunately' our standard of accepting proposals as the truth even though we know they have been modified and, therefore, as such, not the truth anymore.

However, this book is purported to depict the raw reality, the bare honesty to you and me, the bottom-line truth with its spectrum of bitterness and sweetness, with its highs and lows, with its moments of happiness and sadness, with its winnings and losses; the point is this is not your everyday fictional fairytale stories. All of the above are situations inescapable to human life, to be precise, your life and mine are included.

Henceforth, the deliberations discussed in this book are ultimately the truth governed by intellect, objectivity and facts. Consequently, these presentations herewith ought to lead us to achieve a realistic prospect with which you can opt to activate your 'immortal intelligence' should you choose to do so. But essentially the decision is under your free will.

As such, you can imagine that it has not been my preference to even write introductions in essays for the sake of writing introductions or following their routine. I am rather a 'get to the point guy'; henceforth, I make herewith a disclaimer of my unorthodox approach to my writing style. As far as I am concerned, writing is a method of communicating a concept to another or an audience, for that matter. Therefore, as such, the sooner I can deliver the substantive point to its addressee, ironically the better with the least amount of words and explanations.

That is because as much as we set for ourselves to work on goals in order to reach our aims and benchmarks, yet all the while without any let-ups, we remain amongst the 'obliged', which means we have to earn a living for our existence and that requires time and energy, coupled with effort and dedication to sometimes exhaustive levels.

Because the essence of life makes us obliged to carry out our duties for our existence, there is no telling; therefore, the need to balance your immediate requirement for your survival with the major aims and achievements for the long-term targets you may have. The least I can say is good luck with this journey. That said though, with a caveat, being do not achieve one at the expense of the other.

By that, it means do not allow your efforts for survival in life, 'paying bills', etcetera, to exterminate your longing to achieve your higher targets. Let me make myself an example ;as I am writing this paragraph, believe me, I can be somewhere else earning 'incremental' income, which can be a contributory means towards running costs 'bills' and other requirements such as going on a vacation and the like. However, in order to reach our major goals out of life, we need to sacrifice some of our minor gains. In my case, I am not sure whether this first piece of professional writing of mine will ever see the light of the day or even attract one reader and yet I have to try with my best efforts relative to my circumstances in life.

Firstly, let us learn together that there should not be anything too complicated for the rational 'sane' average human mind, and henceforth, phrases such as 'it is too complicated' or 'too deep' are mechanisms employed to freeze your inquisitive mind to rise higher or even reach its potential. In most cases, the owner of these phrases utters such words in order to mask his/her intentions or even refrain you from seeing the shortcomings of his/her 'flaw' or even deception in the anatomy of whatever product/service that they are selling.

Other times, the utterer has defective communicational skills to elucidate what they exactly mean. The real redress 'solution' to this is never to buy this 'notion' again in your life and engrain yourself with it. If something 'concept' is too complicated or too deep for human beings, that is who we are after all, then it ought not to be offered to human beings in the first place. So let them pitch such a 'complicated' concept to the aliens or whatever their audience may be because certainly, if they say it is too complicated, then it is not for you and me.

For example, many national banks inject printed money into the economy via the retail high street banks, which, in turn, is supposed to circulate into the common folk's pockets and the markets. This process could sound to you and me a complicated matter when we hear its conceptual pronoun such as 'quantitative easing', but it should not be for the sane average human mind. All that is saying is the Federal Reserve (US) and National Banks such as Bank of England create digital money which then is printed as extra cash at what value? The simple answer is naught.

It is so true that you will own one life experience on planet Earth; thus, it is paramount to make it count, notwithstanding any direction you embark. This is so reality as a substance of fact because there is no sensical human who can claim this is my 'third lifetime' 'reincarnation' on planet Earth. Last time, I was a good 'upright' cat to get to this promotion despite there being some mythical claimers out there for this doctrine. In other words, this is one shot in life, so what do you want to do with it is the climactic question for any functioning intelligence, and if you think about it, there are solely two outcomes: either you make it count or waste it, and fundamentally there is no third option.

A parable is given: Just imagine you have been allowed to eat from your favorite restaurant once in your lifetime. What would you order? Would you go for quality 'long-term benefit' or rather momentary enjoyment from the Menu? Would you opt for healthy steamed salmon fish with broccoli and drink with nutritious coconut water or go for high cholesterol spiced fried chicken and a whole bottle of hundred-years-old Scotch to be wasted and get the sensation of intoxication with its inevitable hangover bangs the next day? Would it be the intellect of the mind or the pulse of the heart that would dictate your choice of actions?

I am sure there are some other choices in the Menu, nevertheless, this was to illustrate whatever choices we make come not only with sacrifice but consequences too. As the one who chose the former is a long-term thinker and the one who chose the latter is a momentary enjoyer, therefore, which camp do you want to join? Who are you more of the time because certainly there is no 'absolutism' with any one of us, nor shall we expect that anyway.

As the title of the book suggests, the optional need to activate your immortal intelligence! A clear presupposition has been passed, which projects that you already possess perpetual intelligence which can be transcendent to 'time and space'. However, the question is, have you engaged with it yet? Have you utilized your powerful mental tools and faculties, which are capable of alleviating the constraints of time limitations such as ageing, decaying, and ultimately coming to our demise?

Instead of attempting to hide, deny and flee from this ultimate reality called death, we will engage with it and interpret it to mean new opportunity, yes, you heard that right: death 'itself' can be viewed as a new opportunity, new life and existence. Even we should include our perspective, thinking of it somewhat enjoyment, pleasurable and delightful. However, only through activating your immortal intelligence you can reach that level. Why would you want to settle for anything less than that? A parable is rendered; one who neglects this opportunity of a lifetime would be indifferent to one who let go of an opportunity to make a million dollars only to focus on making a few thousand dollars. The beauty of this is there is no either/or; you do not have to sacrifice a few thousand dollars in order to gain the million, which means you may aspire to achieve it all; why not?

Any subservient ambition amounts to 'letting yourself down'. Your full potential intelligence has been deactivated as it exclusively only engaged with that which is temporary, that which is short and definitely that which is mortal. We should neither aim at mortality, demise, decaying and death nor lead a pretentious life span which you already know its final end.

At best, the mortal life will deliver for you a good funeral and a brief time of eulogy with a couple more tears along with hopeful good wishes. Not only is it neglectful, but it is regretful of colossal proportions to settle with the above when you know you can achieve perpetuity of gratifications through activating immortal intelligence, which is astonishingly something most of us already possess.

What is Intelligence?

The main body of this book will not be consumed on whether we possess intelligence or not, as this should be widely known 'a priori' to most of us, but rather whether we are making it count for our goals to be successful will be the topic of our discussion. The pronoun human being 'in and of itself' gives the provision of intelligence embedded with it.

This characteristic is inherent within the fabric of human ingredients, for which reason humans are distinct from the rest of the animal world. However, this notion is not as clear as we might think due to the spectrum of behavior of human beings with respect to the animal world. A human whose greatest ambition revolves around attaining momentary excitement may not be far different from the lives of many animals, or it is even plausible the two could be equated with each other, with the exception of some natural anomalies.

If you look at it from one perspective, some animals seek good land to graze on or hunt; the more fertile the land, the more preferable. This equates with a human seeking the highest-paid job in the employment market or business; both are effectively searching for the best quality provisions attainable not only for themselves but, most of the time, for their families as well. The similarities carry on in many other aspects within the behavior of species, overarching rivalry on resources, including that valuable spouse, while for animals they compete for mating rights; for humans, they seek that soul-mate love in partnership and companionship in many regards with its own undeclared competition as well as rivalry.

Dual Mechanism of Intelligence

Let us herewith peel off intelligence to its bare skeleton and, henceforth, classify its dual mechanisms of offensive and defensive divisions in nature and approach. In simple terms, the defensive division of intelligence is supposed to protect you from harm, while the offensive division of intelligence is supposed to gain you a benefit. The higher the stake of both divisions, the more efficacy of intelligence you ought to utilize for that situation and circumstances.

An example is given, let us assume you are a self-employed commission-based worker; hence, the intelligence utilized in order to discharge your duties while working, which actually generates income for your gain, is the offensive division, hence gaining the benefit of the wages/money, etcetera. However, it is imperative to get to work and come back safely; thus, if you commute to work via driving, the intelligence utilized to safely complete the journey is the defensive division. Otherwise, you may lose all the benefits intended to be gained altogether if, God forbid, you were to crash on your way to work.

Consequently, these dual mechanisms of defensive and offensive divisions play a pivotal role in achieving your success. Furthermore, what is oddly interesting is the fact that almost all of our failures are due to our own doing or lack thereof; the root of it would highly likely be traced back to lack of implementing our intelligence efficiently. The truth is, how many times have you heard of someone losing their job due to natural disasters such as earthquakes or hurricanes? Is it not almost none? And that is the fact. Most of the problems we encounter are man-made, whether individually or collectively, and they are the direct failure of not activating the defensive or offensive mechanisms of our intelligence. The degree of the impact of their failures under these divisions varies; one could be more severe than the other.

While the offensive division may leave you destitute because you could not gain benefits 'money' etcetera, the likelihood of failure in the defensive division will 'see to it' your demise. Even fatality is a highly probable destiny; hence, the prioritization of the two ought to be hierarchical, which simply means defense should come first in order to save yourself for another day to try again for the offensive mechanism of intelligence. For what use is it if you are paralyzed irreversibly on your way to work through a car crash?

The principle of 'utilitarianism' is at play, notwithstanding its attribution to Jeremy Bentham. Functional human intelligence has always been calling us to task since primordial 'beginning of time'. Hence, sometimes you need to let go, as we say, 'the small fish' only to catch the 'big fish'.

However, if you know your targeted prize, 'the big fish', will not be able to fit into your net, then as the saying goes, do not bite more than your mouth can chew. In that case, you need to settle with the realistic means which are available to you and tenable to your capability at that particular junction.

Remember, gaining something, however incremental, is always better than ending up with nothing. It is a failure concept to polarize your intentions into the mode of 'all or nothing'. This philosophy is unpragmatic and, many times is the causative component for the loss of many otherwise successful people. That mega achievement you may see in some of the successful people not only is it not a fluke, but it was achieved through the accumulation of incremental and sequential chain of gains.

Thus, the acumen is applying the right wisdom at the right time and place. If one of them is lacking, then the goal may not be achieved. Your thought may be appropriate, while its application may be erroneous; hence, the result will come short of a desirable outcome, or the thought is suitable, but the place is not accommodating.

An example is given: Barack Obama would not likely have been a successful candidate had he run for the presidency of the United States in 1860 and faced off against Abraham Lincoln. The point is any amazing idea must be matched with the right time and place; otherwise, it amounts to any other concept, whether lousy or good, that had previously failed. Now that we have identified the dual mechanisms of intelligence, let us identify the two aims of intelligence: the lower level and the higher level of intelligence. Ironically, it is the failure to comprehend these two levels where our growth as humans beyond other animal species is clogged or even plausibly reversed to shockingly lower than animal species!

The lower level of intelligence, we will call it the 'mortal intelligence', and predictably, it is this one that most of us will ever engage with for the duration of our lifespan. It is as if it is designed that way; it is as though a systemic wave is pushing us that way of thinking. It even feels that we are being denied to even think freely, the very fundamental of all human rights—the right of freedom of thought and conscience. It is indeed as if a monopoly of thought is exercised on us while we are being made 'subjects' under this soft but powerful authority, which could be said is even somewhat dubious.

That is because we are hearing, paradoxically, this is the best timeframe for human freedom and rights, and yet systemically, we are under slavery, deprived of the predicaments which are conducive to making the exercise of free thinking possible, let alone fostering.

As soon as we are born, we are orbiting the same man-made path—we will call it 'the artificial orbit'—which inundates us with a series of suffocating responsibilities, one after the other, until, if we are lucky and reach our geriatric era, even then, we are likely to still be institutionalized all the way to our graves.

The reason why the word "orbiting" is selected as a descriptive mechanism of our lives is the fact that the path designed for us to go through during our lifespan lacks thereof any alternatives.

We are told through soft but powerful systemic control what to think and what not to think, what to preoccupy our intelligence with and chase after a never-ending 'dream'. If you are one of the fortunate ones who achieve 'the fame and fortune' , then it requires equal if not more of your attention and intelligence to uphold and maintain the status quo, because suddenly you might lose it or face predatory hyenas to eat off your kill through civil lawsuits, litigations and everything in between.

The hyenas are notoriously known for their slick tactic of eating off the hard work of other predators, such as leopards in the jungle of Africa, unlike vultures, who at least seem to wait to benefit the corpse when it is disposed of and getting wasted. Surely, there are their human versions in the system which governs mankind, i. e. the industries and institutions, whether it is the stock 'financial' markets, sports, entertainment or politics. Hyenas are always waiting around the corner to devour your kill and leave you hungry and destitute, with no moral due regard towards you. After all, it was your hard work that brought the gain through efficient utilization of the offensive division of your intelligence.

Mortal Intelligence

How does it play out? Before we delve into it, a disclaimer first: this is not accusing the entire system of being rigged in its totality; however, it is to point out, to a large extent, it functions to divert our minds from invoking our higher level of intelligence called the 'immortal intelligence', which will further be deliberated below. Right after our existence begins, we are sent off to an institution, call it 'nursery' or 'kindergarten', which grooms us to our next stage.

There cannot be much blame here; even though we are being groomed, we are still young—off to elementary school.

Perhaps this is the best time of your entire life because you strike the unique equilibrium of happiness, innocence, and play without much consequential responsibility except seldom negligible 'naughty corners'. However, it is here where the seeds of 'mortal intelligence' start; it is time we define what is 'mortal intelligence' as opposed to 'immortal intelligence. ' The 'mortal intelligence' is the lower level of the two intelligences we possess. It is the intelligence whose goal and target dissolve as our lifespan on planet Earth comes to an end. Typically, this means the time, energy, and dedication we put in to acquire a set of skills from which we can live a decent life on planet Earth, nothing more, nothing less.

Your learning curve of both formal and informal education, training, and experience will ever amount to be that said above: only 'skills' from which you can earn a good income until you reach geriatric age. This equates with the animal species out there, animals like lions and killer whales 'Orca'—they, too, learn from their parents how to live a decent life via learning the best hunting skills. It must be said, unfortunately, if this is all your intelligence is prepared to engage with, then irrespective of your fame and amassed wealth, you already lost before you even started.

The truth is, temporary lifespan on planet Earth is so scarce that it is never enough. I, once upon a time, had the opportunity to converse with a 90-year-old rich man who maintained fairly good health and conscience. I recall asking him how life is since you have been living a long and rich life, not only with hedonic pleasures but also the benefits that come with financial freedom.

His answer was as surprising to me as life itself is. He replied, "I wish life begins from top-down as opposed to bottom-top. " By that, he meant, *I wish people were born from 100 years old and started ageing from the opposite direction towards one year old.* From his answer, 'right away, ' I knew life was never enough to a rich man who experienced more financial freedom than the average human being.

I further could not distance myself from the fact that all the skills he mastered in all of his lifetime that took him to the heights rendered themselves redundant as time elapsed. What then for the ordinary person, I wondered? From this, I am certain all the rest of us in lower-income 'class brackets,' whether we admit it or not, would boil down to the same feeling too.

Coming back to our man-made artificial orbiting path, now we are off to a secondary/high school institution. A new kind of pressure is introduced to us, which is whether we are good enough to go to university or not. Sometimes, it is so inflated that the adolescent students develop depressive behaviour or become overwhelmed, which causes them to be neurologically anxious.

This solely purports the accomplishment of 'mortal intelligence', which means supposedly acquiring skills from which to earn a decent income in the future. Any extra time they have, now the urge for entertainment enters into the equation. Therefore, the strategy seems to be the application of experience under 'inflated' suffocating tasks, which gives birth to the need to have an equal release stress button. As 'Newton's law' elucidates, every action calls for equal reaction, and this is at play here in the form of entertainment, which typically develops into the intoxication of the brain that persists into adult life. While the brain is under intoxication, it is divergent from engaging with the higher faculties of the brain— the immortal intelligence.

The variety of operating 'drugs' to achieve this task is innumerable and ever-expansive into innovative and more lethal ways to make the human brain dependent on them. While some of them are openly supported legally in many countries and jurisdictions, such as alcohol and cannabis in many parts of the world, many more are indirectly tolerated, such as cocaine and MDMA. Have you ever seen anybody with money who could not find cocaine or ecstasy in any major city right under the nose of the largest police task forces with their anti-narcotics departments? Let us not put the blame on the police; it is the way the system is designed, which is no coincidence.

The process does not see any remission and is applied potently and consistently throughout college and university years. Except this time, you are likely accumulating debt on top of the stress from the classroom on weekdays, hybrid with becoming wasted in the ballroom on most weekends, even sometimes weekdays too. The brain is engaged with 'mortal intelligence' during the day or under intoxication during the evening, which entails yet the brain is digressed from ever engaging with the bigger picture of the 'immortal intelligence'.

In the prospect that you are not caught with the above divergent mechanisms of drugs and party stimulation, there are plenty of colorful methods out there, even astronomically fancier clothed than we can ever imagine. There are literally uncountable man-made competitions. Supposedly, the best bright minds of Cambridge University students versus Oxford University students need to compete in a traditional boat racing conceived by Charles Merivale of St John's College, Cambridge, in 1829, when he challenged his childhood friend Charles Wordsworth at Christ Church, Oxford. Likewise, the Yale versus Harvard football rivalry and competition was conceived in 1875 for the same effect.

These seemingly innocent recreational competitions require more effort, time, and dedication than the abstract eye meets. The artificial orbit as a phenomenon protracts and reaches every one of us at every level and industry you can fathom. It is embedded and infused into our fundamental human psyche for as long as we are physically and mentally competent. The subject matter continues to civil servants as well. How many times have you come across these types of competitions with an awful lot of preparations, your city police department versus the fire(fighter) department, or even perhaps your city's municipality versus the neighboring city's municipality in games of competition during the weekends?

The goal has been: how can we keep the human brain busy away from ever engaging with the bigger picture during the weekends or their free time because the average human is 'obliged' naturally to earn a living during weekdays and working days, hence would be consumed by the demanding duties from work or school during the weekdays anyway.

Yes, it has been a divergent mechanism. Let us look into the goal behind the commissioning of the so-called the 'beautiful game' football/soccer. Perhaps the largest and most popular competition in the whole world, which makes commanding financial wealth surpassing $4 billion. However, when you factor in the substantive matrix, the English Premier League alone is worth £1. 78 billion, and the football shoe market alone is worth over $18 billion by 2023. These are not only conservative figures but lack the factoring in of other by-products, such as advertisement income as well as live event streaming. Thus, we can project this sport's wealth to shoot into the trillions.

The football game is rooted back to 11th century England under the conquering of Normandy by the French. It is believed to be born out of this unwanted situation for the youth of England. The Normands called it *la soule,* and theoretically, it is imported and commissioned to take the stress off occupied England and perhaps divert the youth from resorting to foster a revolt, as happened in this era after the Battle of Hastings in 1066.

Furthermore, historically, people with power, such as kings or emperors, utilized weekend sports as a vehicle to divert the focus of the masses from holding their administrations to account. Whether it was the Roman Empire using sports entertainment such as the Gladiator Games for that purpose or Nazi Germany investing in the Berlin Olympics of 1936, which was intended to serve the same end, the idea was always the same: the common folk 'peasants' would be struggling with ends meet during the weekdays. Therefore, let us preoccupy their minds during the weekends and free times.

The artificial orbit strengthens this notion vehemently via constructing a flared-up buffet of competitions designed for the common folk. Anything you can and cannot think of is on the menu, and there is a contest for it, so it is best to exhaust and divert their inquisitive intelligence, allocating time, energy and dedication for these activities instead of having alert 'counting' brains.

If you consider the book of Guinness World Records, it is no less than 'absurdity', shocking the magnitude of some of the topics people are being made to compete upon. Meet the champion Miki Sudo of the USA, the 2022 world record holder of the most hotdogs eaten under one minute, or meet the champion Olga Liashchuk from Ukraine, who holds the record of crushing three watermelons between her thighs, these supposedly are to inspire the young generations to come in the future, just to name a few.

If you listen to some of the heated debates and discussions on the subject matter, they include who is the 'GOAT' between Michael Jordan and LeBron James or who is the best actor between Robert De Niro and Al Pacino. While the latter were acting—which means nothing of their doing was substantively reality, although they acted well in their fields—the former and their counterparts 'skillfully' placed a ball into a net, that is all, nothing more, nothing less.

None of these above personalities who occupy the apex of our imaginations have done anything significantly imperative to the development of mankind. It is, therefore, no surprise that they are held and treated more superior to those who substantially impacted the development of mankind. Most of us, including myself, would have no clue or barely remember who found the antibiotics we take against infections and diseases or who contributed to finding the Global Positioning System (GPS) we so conveniently rely on every day when we are commuting to and from work or the like.

We would like to see our children be inspired by people like Dr Alexander Fleming (1881-1955), the founder of antibiotics in 1928, or Roger Lee Easton, Sr. (1921-2014), the physicist who laid the foundation of the system of the global position (GPS). As brilliant minds as they were, notice both are deceased now; both could not create anything to shield or repeal their demise and death irrespective of their competence in applying the best of their 'mortal intelligence'. As we said earlier, this lower intelligence ceases to be any means of benefit to you and me at and after our end on planet Earth.

The manmade orbit path ships us off to what it calls the real-world phase of our life now, as if the past stages were fictitious or did not count towards the ultimate responsibility of our lives, as if death contains a particular reprieve provision for university students, or civil servants and the like. With that said, concurring is reasonable at this juncture due to the fact that the older we get, the more responsible we ought to be.

Now, after university graduation with likely large debt, the orbital path is compelling. You are faced with a miniature income and costly life requirements such as a mortgage and car payments compounded with credit cards and a lifestyle to maintain, not to mention your family and dependents. The road map is sturdily paved before you, and you might as well jump into it because, as daunting as it seems, there are no viable alternatives out there.

As a result, your brain would be locked into grinding through the orbital path so that just maybe after three decades of hard labor and sacrifice—that is, if you survive redundancy, credit-crunch, Covid and other external and internal unforeseen popping-out issues—you might pay off your mortgage, at which point you are ushered into your retirement. All the while, nothing visible to you would ask you to stop for a moment for contemplation and reflection, do some soul searching; quite the opposite, it feels as if the artificial orbit sedated our minds; we were fixated with not defaulting from our obligations financially and otherwise, so much so that we are at some levels behaving like slaves who are under mental occupation.

No doubt, as with everything, there are exceptions; there are the Elon Musks and Richard Bransons in every generation. Even then, in such a 'niche class', happiness is not guaranteed; this thing called happiness, which is the ultimate prize we are all after, cannot be insured for us.

Unlike your products, such as your classic cars, houses and the like, there is no one who can place happiness under warranty. It is not even in the vernacular. That is because all of us, including insurance companies themselves, are under a great deficit for this ultimate trophy. Furthermore, the artificial orbit encourages the 'mountain of mirage' pursuit of wealth generation, that is when a billionaire tells people he/she wants to become a trillionaire.

That is when the pursuit loses its objective value, where the average intellectual human wants money for the value of what it can buy or bring. For instance, the average rational human being may pursue a certain amount of wealth for the value of buying a house, a car or a trip on holiday, etcetera. Therefore, in essence, he/she is looking for not the money itself but the value it yields, whereas the mountain of mirage pursuers—a billionaire—wants to increase his/her wealth so that they can brag about the balance of their accounts or who made it to the top of Forbes this year. To the mountain mirage pursuers, any added money to their wealth will not bring in any additional value which they already did not have the freedom to access.

At its best-valued proposal, the manmade artificial orbit offers the greater purpose of life is the reward of the legacy you shall leave behind. While it seems to be a sensible goal to labor towards, however, this is a perishable product 'in and of itself'. There is a hall-of-famer status to achieve for many sectors, such as Hollywood and sports.

The Hollywood Walk of Fame takes years of serious work, sacrifices and dedication to achieve the star of your name, which is selected in many categories, including professional achievement, longevity in the profession, and contributions to the community. Even then, it is not enough. It undergoes the Hollywood Chamber's Board of Directors subsequent to when and if the Walk of Fame Selection Committee renders its decision of endorsement.

This process astonishingly resembles the dysfunctional stages of the UN General Assembly (GA) and the UN Security Council (SC). You would solely admire the latter only if you were the beneficiary of unjustified 'veto' power and enjoy you can operate outside the law and further deem the rest of the world community to be subservient to the hegemony. Human rights ought not to be classified due to achievements. It should not matter at all if your Nation-State achieves the largest GDP or possesses state-of-the-art military capability.

Human rights ought to transcend the power and glory of a Nation-Stats. It is inherent and universal, which simply entails that because you are born as a human being, you automatically are entitled to those rights irrespective of your residential jurisdiction.

Therefore, regardless of whether your hometown is Gaza or Georgetown, Washington DC, Human Rights applications ought to be indistinguishable 'jus cogens' for the world community. By the way, the hand of God's 'natural selection' had determined which Nation-State or race/ethnicity you would be born into, which means it would amount to be reckless to be a devotee to such above mindset.

For truly many morally responsible humanitarian citizens of the world are wondering how, in this day and age, we are experiencing this high level of 'manmade' human suffering in 2024. What happened to International Law and international legal authorities such as the International Court of Justice (ICJ) and the International Criminal Court (ICC)? Are they clowns who are wearing dark gowns without teeth, or is the system itself rigged?!

What happened to the UN that was established to prevent human suffering after the Holocaust? Why are the civilian population of Gaza facing genocide and ethnic cleansing? Why cannot the people of Ukraine choose for their future and so on and so forth? There is no doubt in the current international legal framework that the Russian invasion of Ukraine, being a sovereign state, would amount to be an illegal invasion for the belligerent state unless Putin can justify it was self-defense, something Bush miserably failed in his Bush doctrine of 'imminent attack' from the Iraq war of 2003 but went ahead with his illegal invasion as though he was the morally righteous leader of the free world anyway.

Furthermore, there is a legal ambiguity that needs to be cleared up with regard to the unprecedented death toll of the people of Gaza, which currently is viewed under the provisions of International Humanitarian Law when Gaza is not a recognized established State. Furthermore, Hamas is not a State authority by the admission of the current Israeli government led by Netanyahu and his far-right Jewish power party coalition ministers, including Itamar Ben-Gvir, who is accused to be an 'illegal settler' himself, as they deem Hamas to be 'a terrorist group'.

Therefore, the recognized State of Israel, whose leaders have 'control' over Gaza and the West Bank 'cannot have their own cake and eat it' by conducting grave violations of human rights in crimes against humanity and further not allowing any impartial and independent international investigation. These 'alleged' violations include; grave inhumane acts of suffering, genocide, torture, ethnic cleansing, persecution and extermination as such, therefore should be found applicable and liable under International Human Rights Law as well.

The inhabitants of territories such as Gaza and the West Bank are under the control of the Israeli State, which has complete control over their borders, taxes, GDP, etcetera 'all the signs of

State' as a result it gives rise to the fact that Israeli State owes 'positive' obligations and duties to protect and safeguard the human rights of the Palestinian ethnicity/race under their authority.

Therefore, Palestinians residing within these territories under the control of the State of Israel ought to have been granted the same rights as the rest of Israeli State citizens, notwithstanding their expression and aspiration to be an independent State sometime in the future, which legally has no ground to be used against the Palestinian ethnicity/race under the authority of the State of Israel since the ambition of 'self-determination' on the part of the Palestinians remain a conceptual goal in the mind at this juncture.

The principle of the law, therefore, ought to apply to the substantive reality on the ground rather than the mere symbolic thoughtful ambitions of the Palestinian people. Quite paradoxically, the Israeli State can only invoke the sole International Humanitarian Law if the desired Palestinian State is established and functioning. However, anything less than that, such as the current status quo, would qualify every innocent civilian who died in Gaza in a hospital shelling, school bombing and the like, such as induced starvation.

Every UN aid worker and Press journalist who were killed with such an intent while carrying out their humanitarian duties in this violence, their suffering, blood and death ought not to be justified under International Humanitarian Law as 'a collateral damage' for a legitimate target because Gaza has been fundamentally under the control of the State of Israel. Consequently, the leaders of such a

State must treat the civilians under their control as the rest of their fellow citizens in the State until such time the territories in question evolve to become a full-fledged State which has full control of its borders, tax and GDP at which time the sole applicability of International Humanitarian Law could have the legal precedence over the International Human Rights Law. Even then, it would amount to violation to recklessly aim at prohibited and civilian installations such as UN and press stations and personnel, furthermore, the 'principle of proportionality' would bar killing large numbers of the civilian population, as witnessed in Gaza.

It also 'may' mean that you as a citizen 'recklessly' participated in such grave violations of human rights, provided you elected your government and did not hold it accountable during the election in relation to the subject matter. Is it not a grave contradiction to hear from our governments that we have to 'cut back' in order to be responsible for the environment and the preservation of animal life, be it polar bears or others and yet still they participate in the hefty sales and shipment of weaponry and ammunition used in conflict zones. How can one claim to care about polar bears over real people, with the misleading claim to carry out those lucrative military sales in the name of 'defense' when the conflict is local to another continent and places far away?

Is it not ironic that the planet Earth is more accommodative to all its inhabitants multiple times over, its landmass and/or resources more than suffice the whole global population multiple folds over, and yet still, due to the genocide in Gaza, our news outlets are no short of watching horror movies even worse than the 'Evil Dead(1981)'.

Except this is real life; people, families, children and mothers, amongst others such as elders and so on, are bloodied and killed with the illegal 'bombs' falling from the sky like the 'purple rain' of Prince. Except this time, each drop is red-blooded rain soaked in real human suffering and tragic stories under our watch coupled with shrapnels precipitating like the breeze of air, except this time like unhinged category five hurricane sparing none, not even an iota wiping off entire families like mass 'Chinese style' spraying Covid; except this time it is not 'a virus' that is the target but real human beings like ourselves and nothing is out of the range of shooting aim as if international human rights and humanitarian law protections are decorative jewellery one wears at will whether be it schools, hospitals, religious temples, places where the international law declared to be safe haven for the civilian population.

Yet we, those of us who hope to be sensible and humanitarian, appease our morally responsible hearts that this is inflicted by the 'decision makers' at the top of the government, not us, the ordinary society, completely delusional!

It was our collective decision on election day that gave these blood-swimming politicians their mandate on our behalf in the first place in a democracy, which we maintain its oversight when their term comes to an end. This is compounded by the fact that it is highly likely that our governments are not only shipping weapons without our consent in our name but also the tax citizens pay while the citizens are lacking some of the basic essential services such as good hospitals and health services, peaceful schools and education to young children and adolescence along with job creation for the communities at large.

Our civil servants' overarching teachers, nurses, firefighters are complainants of this manmade inflation, which is the result of egos and warfare profiteering.

It is time each one of us without distinction stands in front of the mirror, only this time to seriously look into our souls and evaluate our morals, not just to check out our make-ups and the designer clothes we want to wear as the 'artificial orbit' wishes to demote our intelligence.

You are not only responsible for a global environmental footprint but more preciously global 'peace' footprint by doing your part, however 'incremental'.

Comprehension is granted where there is genuine and proportional self-defense. Ultimately, however, the greatest collective failure of humanity as a whole on the planet Earth will be our negligence in keeping and maintaining world peace. As said previously, the planet Earth can meet our life needs and demands amply. The current world order encompassing the UN and the International Justice systems being hierarchal and hegemonic can no longer justify its validity except through power and fear, which is the antithesis to legitimacy and justification, where the alpha leader suppresses his subjects without the rule of law, justice and equality.

Only through his virtue of power, be it 'veto' in the security council only lasts until when he/she is overthrown by another power, however unlikely it seems, which is the inevitable end of that status quo. I hope to see a world transcendent to the above scenario playing out. I hope to see a world where everyone can access justice and rights equally.

Coming back to the double authentication of Hollywood granting you their immortal version of 'Hall of Fame' After the rigorous process of selecting you to be promoted with your star, your name would find itself on the streets of Hollywood as a walk of fame 'star', only people to 'piss on' after night outs, you can join the fray pets as well such as dogs and cats.

Additionally, some drug-addict 'junkies' shelter on many stars, as reported these days in Hollywood. A functioning intellect must wonder, "Is this 'the legacy?", "Is this the insurance I was sold?" This is no less than reputation bankruptcy. Furthermore, time will eclipse your fame. The simple truth is people, if at all, seldom recognize you. I mean, how many of the contemporaries of today's generation remember James Cagney, the best Hollywood actor of his day 'Angels with Dirty Faces (1938)' or Humphrey Bogart of 'Casablanca (1942)' both were the greatest achievers in the filming industry. Ironically, if the 'artificial orbit' convinces you somehow that you can outdo these personalities, if you accept this proposal, you can call yourself smart, but history has already proven you as a hallucinator. As the saying goes, 'history repeats itself'.

Even those who have really achieved technological advancement that impacted human development are not immuned from time eclipsing their legacy as well. If we were to be quizzed on who founded the 'telegraph', which was a means of global telecommunication for a long time would be strange to us, even news altogether, myself included. Mr Samuel Morse, in 1830, revolutionized the world and contributed to the actuation of 'globalization'. It is expected your brain is saying to you these people are ancient, but we need to fathom that 'today is the yesterday of tomorrow'.

The point is the world would move on from you and me. We would be forgotten, even oblivion from the brains of the next generations. There is no legacy in the artificial orbit lifestyle proposal, regardless of any stature you might reach within it. The fact is, it is an illusion, the Mark Zuckerberg and the Bill Gates of today would wither away into the fogs and winds of the time as those before them.

It was in the summer of July 1969 that astronaut Neil Armstrong (1930-2012) and his team conquered the moon, the world held its breath, human imagination, technology and simply pride went up to its all-time high, President Kennedy's (1917-1963) daring speech of we will conquer 'the moon' in September 1962 was delivered as well-deserved spoils of war, it was celebrated like the Alexander the Great's majestic conquering celebrations of the past.

Nevertheless, none of these great men could extend youthfulness to themselves or shield the inevitable demise and death from themselves as dreamers as they have been; they were caught with the artificial orbit's mirage offerings. This whole saga of the artificial orbit lifestyle and its illusionary goals culminate in the above bleak end. Consequently, for the functioning great intelligence, it is time to re-evaluate; it is time to go back to the drawing board, reassess the foundations of what is life's blueprint and come up with a truthful and objective journey to test it, to scrutinize it; whether there is something beyond this illusionary path or not, which is the topic that we will encounter next 'the immortal intelligence. '

Therefore, this strategy of not allowing the human brain to engage with the bigger picture, 'the immortal intelligence, ' is applied upon us as an orbit in which we have to do our motion back and forth just like a clock's oscillating motion. If you are the oscillator, you did not go anywhere due to your motion of back and forth in the same distance while the clock 'time' over you has already moved on.

Consequently, it is likely we are in between either 'the action' of the stress of school and work duties and 'the reaction' of developing the addictive remedy of intoxicating our minds so that our free time is also wasted. Or alternatively, you are engaged in some other activity which does not engage with the higher cause of our existence. If you find yourself in such a situation, then realize you are being diverted from activating your higher intelligence, your 'immortal intelligence', the never-dying thinking style and ambition. Now it time to define what is this concept of 'immortal intelligence.'

Immortal Intelligence

As the name suggests, this is never decaying intelligence. This is an everlasting mindset. However, before we intellectually engage with this fascinating topic, let us temporarily revert to what we have covered thus far. What does it mean in relation to our topic at hand, the 'immortal intelligence'? All that we have discussed above was directly addressed to our lower-end intelligence; we called it the 'mortal intelligence'.

We said this is arguably all the intelligence we would ever activate and apply during our lifetime on planet Earth. This lower intelligence, however, plays a crucial role in taking us from point A to point B meaning attaining short-term goals. In other words, we called it the 'skills intelligence'. Most of our efforts, learning and training, boil down to obtaining 'skills' from which we can live a comfortable life, a decent life, a self-fulfilling life, a dream life, whatever you want to call it while we are on the planet Earth prior to the inevitable realization of death.

These skills overarch from lawyers, accountants, doctors, teachers or fancy skills such as catwalk models, Formula One drivers, boxers, professional tennis players, film star actors and everything in between. The reason why this is the case is because of its immediacy and proximity to our literal vision and impact compounded by the consistent pressure-feed created by what we called the manmade artificial orbit path, which acts upon us as a chain of pre-occupational challenges to our minds thereby acting as a blocking levy until we reach the inevitable and non-reversible end of our lifespan 'death' without ever thoroughly scrutinizing for what purpose we have been brought into existence on the planet Earth.

The proceeds of the 'artificial orbit's lifestyle are not only immediate and connected to our plain eyesight directly, but they are necessary to our survival as decent human beings in our communities. Recall we talked about the concept of 'obliged', which simply meant we have to pay bills for us to exist and flourish, but we questioned to what extent, at the expense of the big question, are we engaged with that.

Is that all we are about, all we will ever amount to be, to the dismay of the artificial orbit system or any other entity, whether it is your atheist professor, religious leader, the Pope, the Imam, the Priest, the Swamist, the Dalia Lama, the Gyani or the Ayatullah, the organized 'clergy', or any other 'so-called' authority without exception including your spouse and director at work. Not even your parents or children unless you are underage yourself, an insane or mentally frail elderly person, in which case you qualify to be 'non-obliged', which means you are non-responsible for your actions; all the rest of you sane and capable people ought to be able to set aside for a moment all doctrines, concepts and theological faiths infused in you while you were a child in the culture you were born into.

The 'Immortal Intelligence' is the higher end of our brain, transcendent to just being busy with finding entertainment and career-related efforts. Despite their importance, it recognizes that nevertheless they are temporary and expire as we reach our retirement age 'if we reach' that we can no longer carry out or at death, the immortal intelligence is not satisfactorily thrilled with 45 years of career will be the best that life can offer.

The immortal intelligence seeks deeper and more meaningful answers to our existence and stays clever to dodge the divergent(s) and time wasters that the artificial orbit throws at it. Further, it dictates that this search pertinent to making sense of the 'purpose of life' must be brutal and selfish. It must be tailored to you, with everything else being a candidate to be dispensable should they stand in the way. It is absolutely necessary to be this way because no one will die for you or with you for that matter; no one will take the bangs of death for you, nor want to lower themselves that dark, scary 'grave' for you.

If you are thinking about 'cremation, ' it is even worse; no one wants to be in the 'self-inflicted' fire for you, even if they extend sympathy towards you, i. e. family and friends, possibly strangers too; their support is limited to in so far as that emotional discharge towards you.

Due to this brutal reality, this journey should be selfishly individualistic in nature to the last drop of this word like nothing else you have ever engaged in your lifetime. You cannot afford to delegate this quest to someone else, just for the same reason you cannot transfer your death to someone else.

If you want to test this, go ask your religious leader or anybody who is your role model. Can you die for me? Give me your remaining life span so that I can live longer! Their answer is resounding 'no' even if they say yes to you. How do you know this? Just reverse the question to yourself: is there anyone already dead that you want to trade your life with so that you take their status of death?

The answer is yet again the same firm: 'no' unless you have a mental illness and are already 'suicidal', in which case this discussion is not for you, please seek professional help. Death is a non-transferable duty so as to learn its objective and proceed; why is it that death is the inevitable end for us? To unravel this mystery, you can no longer just blindly rely on the indoctrination of others, even if it seems sweet.

Remember all the scammers narrate sweet stories to the ears, soothing music to the heart, the touchy poem to the ego or lucrative wealth to your desires before you find out that you have been scammed, in this case after 'death' too late, which makes the topic all the more climactic, there is 'almost' no room for error for this one. If it is too good to be true, then it is definitely too good to be true. I am sure you must have heard this before; it has not been said in vain; it simply has a historical and factual reoccurrence.

Further, do not get caught in procrastinating and delaying this quest, thinking that you have somehow time, that heaven rendered an amnesty for you, perhaps you are young or rich or lucky. All you need to do is just look around you, how many young and famous people are gone too early the so-called 'untimely departure'.

The young actor Heath Ledger died at 27 years old. If you think many of these rich Hollywood actors' deaths are their own doing by overdosing their bodies with drugs, how about the famous Princess Diana at her prime age of 36 and, yes, a drugs-free lady? The point is there are no guarantees or respite from our inevitable demise. In fact, it is time we give meaning to this concept known as death and perhaps remove the shackles of its taboo so that we face it 'head to head' since we have a definite and unmissable appointment with it sooner or later.

Please note, however, we will engage with this topic further down the line in a substantive way. However, for now, death is a transition, death is a motion, and death is a movement from one place to another. Hence, death occurs when life moves from the planet Earth to its next stage. Even if it is too 'uncertain' or 'unsettled' for you to take this in, just remember death is something superior to our control and consequently ends our presence on planet Earth with or without our consent.

It overrides the free-will we are accustomed to hearing is our inherent right pragmatically; it dismantles all of the aforementioned manmade artificial orbit dreams and their proceeds altogether. It is meaningless at the junction of death how many classic cars Jay Leno accumulated in his garage in California, or how many properties a tycoon hoarded in his portfolio, how many Oscars and Grammys a star received, Presidential Medal of Freedom, OBE from the British monarch, or Sirs and Dames for that matter. It does not matter if the king of Saudi Arabia placed on your neck that gigantic golden chain and sword or even if you are the King yourself.

The odd result of our departure may give rise to, in many cases, that our testator properties we worked so hard 'sweat and bled' and left behind may be distributed as spoils of war, surely someone else will enjoy them, hopefully, our offspring. Other times, if you were lucky and famous, your memorabilia would be auctioned. This typically happens because even our loved ones do no longer have the hard stomach to watch us as a dead 'legend' any longer because they want to move on, and also because your demise reignites that loss and mourning feelings culminated with the artificial orbit's pushing us to distance ourselves as far away as possible from engaging with our ultimate home of death.

With this absolute and direct power 'death' exerts over us is also assimilated with ageing 'itself'. Yes, ageing, too, is a scaled-down death occurring progressively in our body; it transports this observable fact with a similar effect on us. We do not have any capability to shield ourselves from it, just like death, contrary to common belief.

Perhaps, with good health practices in the combination of a healthy lifestyle of accessing good food, a healthy sleeping pattern as well as sporty exercises, we can seemingly achieve a mild 'anti-ageing' capability, which in layman's terms means the hope to only postpone the degree of our cells dying and decaying indefinitely over time, together with other bodily losses such as tissue and organisms due to this unwavering motion of ageing, not age-reversing what (so)ever.

As a result, do not be caught under the artificial orbit's rhetoric in this. Please note the above cell death is not the periodic regenerative cell replacement the human body 'involuntarily' conducts, better termed as programmed cell death (PCD), in order to modulate surplus and/or unneeded cells coupled with recovery mode activation, 'Apoptosis' and 'Nercosis'.

Nothing in the spectrum of priorities can even be remotely equated with the 'utmost' significance of engaging with and making sense of our existence on planet Earth, the object of our being, not only ourselves as the apex creatures in the hierarchical being on the planet but simultaneously the environment around us; everything in existence, from the birds flying in our air to the fish swimming under the salty and sweet waters, are shouting at us with an intricacy of sophistication and awe that one needs to wander and ponder over.

From the Amur Falcon bird that migrates from Siberian harsh winter, transiting through India, to the livability of African weather, only to reverse the journey a year later with natural built-in GPS, to the sea turtles utilizing the usage of planet Earth's magnetic field as natural GPS to navigate in their epic transocean migrations for their 'natal homing'; the marveling process of returning to their birthplace with the vast ocean with no recognizable marks.

Furthermore, if we observe the ergonomics of our anatomy alone, it leaves us scratching our heads. Whoever designed humans to have fingers enabling them to hold a pen and/or type on a laptop, among other disciplines, whilst concurrently equipping the feline 'big cats, ' including lions, with natural night vision technology conducive to enhanced sight in the dark of the night; a technology for humans in night vision devices (NVD) which transforms the photons into electrons and subsequently increases the electrons for better vision predicated on the discovery of physicist Kalman Tihanyi of 'infrared' with its first purposeful use in the military in 1939, the so-called Second World War (SWW).

Bear in mind all the while, human existence goes back so far as known, 300-thousand years ago, based on the science of today. This is, coupled with the big cat's claws 'built-in knife' to tear the hard skin of a buffalo that can withstand all weather elements without needing a further coat, is no short of a spectacle to behold.

The ergonomics of our human fingers, which we took 'for granted' unilaterally, can point to us that our designer and maker

41

intended for us far more intellectually advanced tasks than our counterparts with claws and natural night visions. These must include the ability to write, which means also the ability to read. This, then, is literacy and, therefore, translates to being knowledgeable of concepts, not only through our personal experiences but through the experiences of others.

Hence, knowledge can be recorded and be 'cumulative' in due course. This also entails that through the pen, we are growing both individually and collectively as we ought to pick up from where the generation before us left it off, hence clearly making us a league above the rest of the vast creatures that share with us the planet Earth. The astonishing part is this superiority that we unilaterally enjoy over all other species without any exception is literally granted to us as a means of a handed donation, consequently not earned in any way what (so)ever.

These 'privileges' which are granted to us via 'donation, ' which means we, as its 'recipients, ' in order to attain are not required to trade them with anything else, neither our labor nor our money, that is because its supplier is 'omnificent, ' which means there is nothing we can offer for his advancement or enjoyment, unlike the labor we give to our employer who directly benefits from the proceeds of such labor or service.

The profit may buy them some benefit, whether it is a new mansion, yacht, private plane, or any other material they need, in the sense of without profiteering from the labor given, they would not be able to acquire these accessories, which makes the few privileged owners remain 'needers' of external income to carry out any of their transactions, even when it is not essential for survival.

These great 'donations' begin with the most essential and necessary for human life to exist, which is none other than the 'oxygen' we need to inhale. Without oxygen, the average human being would cease to exist in six minutes, without water in three to four days, and for good measure, without energy, electricity, oil, etc.; it is not even a qualification for life's existential requirements, as mankind had lived millennia without them in the past.

While the Creator retained the supply of oxygen to the general population and thereby provided for free of charge without any recognition for it through the process of 'the oxygen cycle', the middle-man who supplies the energy we consume in our houses generates an average revenue of $1. 8 billion a day in the USA alone; you can imagine what that holds for global energy supply.

We know the fact that many middle and low-income families are struggling with energy bills due to many man-made price hikers, including offensive wars in the name of 'defense' and fears of conflict, which create 'rush hour' demand for the weapon industry. Wars like the Ukraine war or the Middle Eastern instability, the India versus Pakistan fear, Saudi Arabia versus Iran and its proxies' fear, Brexit and everything in-between, causing many of these families to choose between 'heat or eat, ' especially during the winter months.

Just imagine the supply of oxygen would be commissioned to these giant profit-oriented companies. How would that dent the average family's income? 'Unfathomable.' Thanks to the retention of the Creator for the utmost necessity of them all, the supply of 'oxygen' is retained through the 'oxygen gas cycle system, ' which bars any middle-man, be it these powerful giant energy suppliers we know today, from monopolizing its vital supply for the existence of mankind. I can only imagine the evil oppressors and tyrants, if they were to have control over the supply of oxygen, they would be weaponizing it for genocide, ethnic cleansing, and crimes against humanity, and you would wish you were not on the receiving end of them, not even the opposition party leader in a democracy.

The activation of immortal intelligence requires that we do not take any concept or doctrine for granted. Hence, every philosophy or doctrine, be it a scientific theory or religious theology, must be tested, scrutinized, and ultimately, it must pass our logic and reasoning. In other words, it should simply make sense to our intellectual faculties; otherwise, it must fail.

Notice is granted to those of you who are wondering how can the two be reconciled when religion is supposed to be faith, and immortal intelligence requires reasoning and evidence; one needs to notice the proposal of 'blind faith' in and of itself is telling you this belief system does not make sense. This particular religion will fail on multifarious aspects of human reasoning, be it science, historical accuracy, logic, and reasoning. This admittance of its initiation is designed to bribe our intellectual reasoning, thereby sedating your thinking capability.

It is understandable when you are a very young kid, but immortal intelligence does not and should not be expected to swallow this unknown capsule with its fundamental flaw in the name of faith.

No one bases blind faith in his/her most important goals in life. Why is this any different, especially when the issue at hand is a candidate to being the most important endeavor we will ever take? It was not blind faith that made you choose your spouse, your college degree, your career, your car, your mortgage, your gadgets, even your fragrance and fashion clothes, etcetera. An example is granted: when you were deciding on which mobile/cellular phone to buy, you weighed up multiple factors, you called to task your intellectual faculties, and you probably deliberated the most updated phone in tech, the most reliable with a good brand name matching the price budget you can afford for this endeavor.

Astonishingly, a lot of you may have utilized all the ways and methodologies available to them, such as comparing and contrasting with gadget comparison websites. Many more even went to stores to ask hard questions, which is a type of 'scrutinizing' the seller. The shocking scenario would be if there is a judgment day with 'God' being the judge and asks what prevented you from finding me? How would we justify this? Yes, we investigated everything from the gadgets we bought to which shop had sales this week.

However, when it comes to the 'purpose of life', we followed blind faith, that is if your position to this question did not come about via exhaustive 'deductive' reasoning. At that moment, you already know you let yourself down, let your intelligence down. However, it is not all bad news; if somehow you find yourself reading this book, you should embark on the most epic intellectual journey of your life, which is without a doubt to make sense of 'why are you here' and how this universe came about? If any, who is the possible candidate that possesses such a capability to be the source behind the universe?

It has been documented most of mankind would fall into one of these categories, that is, you are likely to be a member of these historical tents: a) 'Beliefenism', which for the most part means I blindly follow some kind of god and religion due to my culture or popularism, b) 'Atheism' which relies on the presumption that there is no God at all, there is no source behind the creation c) 'Agnosticism' which entails my intellect acknowledges the existence of a source of power behind the universe, but I cannot ascertain who it is. This is compounded by further developing positions about the subject matter, which include d) 'Neutralism'- this category relies on the claim that there is no empirical research or evidence to tilt me in any direction thus far. Because of that, I hold a neutral position about the subject matter, e) 'Selfenism'- which entails I am my own god, I know right from wrong, I am good enough for myself, and finally, f) 'Deductionism' - which entails whatever decision I hold it came through deduction of reasoning and observation of facts and science combined with the logic of probabilities.

There is, of course, only one possible correct answer from the above positions, which arises from the processes of engaging with them. As we attempt to put each ideology to the test and see if it can stand erect on the basis of science, historical accuracy, logic, and reasoning and, ultimately, if it can make sense to our immortal intelligence.

However, the tests are not exhaustive in nature themselves 'otherwise', we would be here all day 'so to speak', but are enough to form a foundational understanding to base on our free-will decision to reach the deductive truth as this free-will will be revoked at our death; at which point, it has been withdrawn and can no longer hold the ability to carry out this kind of all-important research to make sense of your existence on planet Earth.

The Immortal Intelligence
Engages with Selfanism

The idea of 'Selfanism', which entails the achievement of self-reliance, one's dependence is only upon him/herself, is important in life. It leads to the feeling of self-dignity and gratitude. It encourages the 'selfener' to seek to be financially independent and to drive him/herself hard to accomplish his/her goals. We owe some of our technological advancements to this very idea.

Many of great minds and entrepreneurs were 'selfeners'; the likes of Elon Musk and Steve Jobs embodied themselves with 'selfanism'. I do not need to stay on the status quo and be adherent to the old way of doing things. Hence, they relied on themselves to achieve daring and new territories of human consumption. In the case of Elon Musk, the world would pay tribute to him for the possibility of mainstream automotive electrification. He relied on his findings, trial and error to achieve his goal while the major automotive companies such as GM and VW thought this was not possible, which is the reason why Tesla Inc gained so quickly the world's biggest market capitalization at $565 billion, while for instance, BMW is sitting at $66 billion, despite Tesla being only founded in 2003 whereas BMW was founded in 1916.

Steve Jobs, on the other hand, with his long list of achievements such as computer and mouse graphics, changed the way we interact with each other with the iPhone in 2007, which completely ignited the boom of the smartphone; one gadget that can do everything from banking to shopping and emailing to colleagues. There are a plethora of materials covering this topic out there should you be interested in expanding your grip on this subject matter. However, despite the above gains being beneficial in so far as mortal intelligence, which we said is temporary and dies with our demise, as such, without any disrespect, it is fair to say Mr Steve Jobs will never use the iPhone again, which deflates the hype and publicity for the skills intelligence. Not one of us can claim to be our own 'God' because no one of us knows how their demise will come about; what will be our cause of death? Is it an accident, cancer, or stroke? The list goes on and on.

Once I met someone claiming to be his own 'God', I asked him a fairly simple question, as I recall: what is the weather like tomorrow so that I can dress accordingly? He automatically checked his smartphone in order to find out the weather forecast. If you need to check the weather forecast for the weather status, there is no point in pretending to be God or, much worse, embarrassing yourself.

If any one of us were our own God, there would not be the need for gender change and the whole confusion of it. If you do not like the gender you have been 'assigned' to, you would snap your fingers and become whatever you wished; in the brutal reality, that is not what is going to happen. Those who intend to medically interfere with their natural gender must go under the knife. If any one of us were our own God, we would probably not be confined within the parameters of this world, sanctioned with its laws of physics, suffering, and pain; we would be in our utopic existences where we would not face the constraints of ageing and death, among various other human limitations.

The concept of secularism in a democracy with its parliamentary supremacy does not make us our own Gods either, notwithstanding making our own laws and decisions of how we will be governed. This, sometimes, is a common misunderstanding with the ordinary folk with narrow and strict interpretations of what is the scope of its enforcement in relation to the 'unseen power' controlling the universe.

An illustration is provided: we as humans autonomously have been doing anything and everything we have been empowered to do as a means of delegation. This concept has not just started with the advent of 'enlightenment' and 'modernity' or democracy but has always existed. Humanity, since the times of 'homo sapiens', had their choice and freedom to decide their food menu and how to dress, etcetera. This should be understood as 'subordinate legislation', which is vested into the power of the executive branch despite the Parliamentary power of supremacy being intact yet still.

The ceiling barrier is if humanity intends to declare 'independence' and implement the literal meaning of self-reliance, i.e., their own 'gods' let us find new gas to depend on for our breathing than the oxygen provided to us; in other words, let us make our own equivalent oxygen or dispose of the requirement of lungs altogether.

The manmade artificial orbit, as said afore, is designed to divert or fool our immortal intelligence to engage with the bigger picture, even if it is an expression as basic as 'oh my God'. Humans always return to their natural disposition as they encounter something frantic or scary, a moment which is the well-established 'oh my God' moment.

There is an intentional campaign to replace 'God' with 'Gosh', and you often hear people saying Gosh in lieu of God. With that said, the paradox is the function of God is beyond a mere pronoun, so much so that erasing its function is unpragmatic. That is why they always end up filling in the empty spot with other names such as Gosh or Mother Nature etcetera, which is a testament that regardless of which angle you look at it from, there is 'an intelligent design' executing plans beyond our human control.

Additionally, another bamboozling strategy is to dilute the meaning of words so that the average human being gets lost in the real meaning and function of words. You must have heard people misuse the word 'create'. I created my children, I created this and that product and the like while we could not create anything whatsoever.

If you put it to the test when we say 'I created you' to our child. What did we create? Did we create the sperm cell or their fertilization, the embryonic development stage, or the fetus? Of course, none of the above is the only answer; the highest point our actions reached is the intention to have a child with our spouse and follow it through the duties of intimacy.

Conversely, the pregnant mother was not even aware of the early stages of embryonic development in her body, let alone creating it. Even when the word is used in a more qualifying way than fertilizing children between spouses, such as when car designers such as Chip Foos or Marcelo Gandini transform their iterations into real, fully functioning vehicles, their creation is restricted to the concept on paper or the clay stage, surely all the raw material employed for this concept was created by 'God' such as the metal, the leather, the plastic etcetera.

Further, this applies to all the significant architectural sights of the world or its prudent engineering, such as one of the world's most significant historical building, Hagia Sophia of Istanbul, with its past association with the world's most infamous theory of indoctrination the 'Trinity' that leaves us astounded to make it work. As well as the fashion cloth we are allured to look at and wear. All of these were incepted via concept followed by enactment and erection.

This is why all that belongs to us in terms of creation is, in fact, 'our intentions'; the idea 'itself', either good or bad, constructing or destructing, truth or lies, that is who we are and what our legacy for the future will be, therefore, in so far as our ability to create is our 'intentions'. We will engage with more of this notion later.

The Immortal Intelligence Engages with Atheism

The primary challenge against atheism has always been the extremely intelligent design and purposeful operation the world around us has displayed throughout time, as already some of it discussed above, which are themselves less than an atom of a regime of concepts which are thought-provoking to our 'immortal intelligence', such as the ergonomics of fingers, the 'un-monopolizing capability' of the supply of oxygen to the masses, and the naturally built-in navigation system of animals.

For instance, that sea turtle with the ability to access the planet's magnetic field for thousands of years is a testament that this is 'an intelligent design'. I hope you have heard of it in the past. Besides that, there is no other rational way to explain but to concur with this concept. If you divert a minority portion of your attention to our human immune system alone, it is no short of a miracle. Let us define what a 'miracle' is. It is when something 'organized' beyond human capability occurs in a manner outside of the human operation.

It is as though this 'intelligent design' is saying to us your capability has limitations, your 'antibiotics' can only be a defense in so far as your bacterial attacks; consequently, you need an external hand as a result, we will provide you with an immune system that repels 'virus' infections. It operates like security, not only catches thieves and peace disturbers but deters them from coming back to the same establishment by keeping their facial recognition images.

Similarly, our immune system keeps a record of every enemy 'microorganism' that has invaded our body that the immune system's radar picked up and defeated so that it can deter it from returning into the body again. This natural immune defensive mechanism where this information is kept is known as the memory cells, which is the white blood cells (B-lymphocytes and T-lymphocytes).

As much as this organized defense system defeats the enemy of 'microorganisms' attempting to trespass onto our bodies, it also discredits the idea of atheism because the concept of atheism argues for an organised order can come from nothing, zero, without author, builder or causer.

This means practically, 'effect' can be without a cause, not to be mistaken with the prospect when we can substantiate the effect but not the cause, which we will elaborate on later in detail. For example, while you and your friend are taking a walk in your neighbourhood on a nice summer evening, you see a beautiful Aston Martin Vanquish parked on the street, and your friend asks who parked this amazing Vanquish here. In your response to the question, even though you may reply, 'I do not know' you never once made the presumption that it parked there by itself, especially as valuable and sophisticated as this supercar is.

Why did you not hold such a presumption? Because you knew full well as a 'matter of fact' that it is unfeasible something as valuable and expensive as a supercar would pop up by itself out of nowhere into the street parking just like that?

It is, therefore, required to activate your 'immortal intelligence' for the same reason. How would it be possible for an immune system which is more complex than a car, to pop up without a cause? Recall in both cases, you could never substantiate the cause behind the effect; you neither saw the owner of the supercar parking it in your neighbourhood street nor the complex cause behind shielding human beings from harmful infections as defense.

It is not, however, the immune system to be preemptive action, but rather, it is designed to repel after the invasion. Please note that the quality of the operation is not being deliberated here, that is, the 'imperfections' of the immune system is well known and evident despite its marvels, otherwise, there would not have been the 'COVID-19 pandemic with its deadly consequences and the like, had the system been bulletproof.

Nevertheless, the defect with the immune system not only does not negate whether there is a designer or not, but frankly, it is irrelevant to the validity of whether a designer exists or not. Just because the car is imperfect and breaks down or even is faulty and goes under a manufacturer recall does not diminish in any way the existence of the car maker, there is always a car maker for every car, whether it is most dependable or least dependable according to your car surveys the likes of J. D. Power. Therefore, the quality of defense the immune system fortifies the body with does not affect the existence of its cause.

Comprehension is granted, though the cause behind the immune system is a higher cause. The term God may be suitable. Therefore, as such, we hold 'God' in higher esteem in 'his' stature than a mere car engineer. As a result, why, if there is a God, his immune system is faulty and not bulletproof at all times? One has to realize one's action and intent would not precisely translate to his capability.

An example is given: a person with a high food appetite of eating would like to restrain their eating habit, the process we know as 'diet'; therefore, as a consequence of their intention, they ate less food even though they could have eaten a larger quantity of food because they chose at that moment to eat less, their action discharged at that particular time may not always be equated with their overall capability. In other words, they could have eaten more but chose to eat less, which also means God had designed it that way.

We knew during the COVID-19 pandemic so many people's immune systems would protect them 'almost' perfectly, and while others succumbed to the virus fairly quickly, in some cases, without clear explanations. While in other cases, there were underlying causes, which were the precipitators to their end and died as a result. The astonishing thing was that there were a group of people who were termed as 'super-spreaders' which, in some cases, their immune systems were so effective that the Virus would enter and get ousted by their immune's defensive mechanisms without them noticing any symptoms.

In other words, they would not be even aware that the virus had entered their body and been defeated and ousted by their robust immune system unless they were tested. Hence, the doctors could not explain this phenomenon at all. The point is this supreme being, 'the intelligent design, ' wills and decrees what it wants, when it wants and how it wants; therefore, the system was created to be 'flawed. '

Yes, it is sophisticated beyond what we can imagine; yes, the immune system does a miraculous job that no doctor, no medicine, no antibiotic can as it shields you and me from millions of viruses such as 'flu viruses' and the like, but nevertheless, it is designed to fail as and when it is 'time' to fail, its imperfection is part of the design of this supreme being 'intelligent design'.

This should not raise any eyebrows. We see our human manufacturers do the same every time with their products. What is even startling is that some of these manufacturers, in their research and development (R&D), are actively seeking methodologies which enable them to shorten the lifespan of their products so that us, as the consumer, come back to buy more of their products; so their goal is profiteering from you and me as much as possible.

It is a complex equilibrium, though, to pull it off because, on the one hand, you want to sell a highly profitable product with a good reputation and, on the other hand, dilute the longevity of the product. The tactic employed in these kinds of products, i. e., the automotive industry in particular, must be superior in every measurement, perhaps for better-term perishable in the near future in areas such as technology, which is time-sensitive and gets 'outdated' fairly quickly, aggregate with, style and fashion of the day, but not other components of longevity.

Additionally, the use of plastic, where the components can be metal, purported to achieve a 'short-lived' lifetime. The immortal intelligence inquires why then the Creator of the immune system renders 'a flawed' system, the answer ought to be obvious so that all human beings experience and taste the process of death.

The only difference is while we as customers pay our hard-earned money to buy deliberately 'flawed' products, the Creator donated to us our immune system; thus, regardless of its imperfect design or not, we cannot bring any rights to claim justice because it was a gift in the first place.

Furthermore, it should be clear by now, therefore, anyone who is subjected to death, that is, anybody who is facing death sooner or later, irrespective of whether they are living or even asserting claims to be gods, which encompasses every living creature on the planet earth, they are qualified to be ruled out to be the cause or Creator of the universe themselves, and their respite of life span may expire any time because no one knows how and when his/her end would come; it could be sudden or gradual but either way inevitable.

With that said above, there is an impressionistic attraction to the idea of atheism, even semi-consensus conduciveness to the immortal intelligence, in the sense that, when and if you are faced with a completely irrational religious belief which its justification is exclusively craned to the human brain via faith, heritage or culture as if it is your first vaccination injection that supposedly protects you from infectious diseases which you despised as a kid.

Many historically significant people were in such a predicament. The patriarch prophet Abraham comes to mind. As reported, young Abraham was born into the house of Idolatries, that is, his father was the carpenter and sculptor of the town's idols. In other words, the manufacturer of these Idols who were the gods people worshipped.

His father wanted to train him as such so that he could take after him. However, young Abraham, with his effective immortal intelligence, could not rationalize that, on the one hand, we manufacture gods while, on the other hand, we are supposed to worship them. To prove his point, he destroyed all the idol gods his father had manufactured in the temple except the biggest one of them, whilst no one saw him.

When people found out what had happened upon their first 'town hall' gathering to discuss, he out-loudly proclaimed it was the big one, the big Idol that did it. I accuse him of this crime. Let us see if he can defend himself. Let the 'defendant' protect himself from this accusation in our lingua franca. In which case, even the

Idol worshippers subconsciously were certain no Idol, whether a god or demigod or a toy, could ever speak.

Then the brilliant young Abraham said, "Exactly if the Idol cannot defend itself from harm, how on earth does it deserve from your thinking intellect to be conferred the title of God unless you deliberately want to 'sell-out' your intellect?"

Please note that Abraham clearly subscribed to singular intelligent and purposeful design and power. Similarly, in some respect, if any young person who intellectually refutes the irrational worship of Idols, pictures, mortals, people, animals, and the like therefore calls him/herself an atheist in the interim, by that, they mean none of these proposals deserves the title of the creator of the universe, then atheism, in this sense, is the superior option than the former which was offered to young Abraham and he rejected.

However, in clarification, not only 'purposeful design' is incompatible with atheism, the immortal intelligence is the antithesis to such a concept as well. Contrary to popular myth, even scientific deduction in conjunction with rationality and reason is established to point out it is impossible that such a measured and proportioned placement of the universe and everything within it can come out of nothing.

Nothing suggests to the immortal human intellect that the universe is coincidental what (so) ever, so as the planet Earth within its unique stance, this scientific position is not novel, far from it. Scientists like Isaac Newton, whom we tribute 'the law of gravitation' along with the technology of 'reflective telescope' says, "Even if there is nothing else, the uniqueness of the human thumb alone would be sufficient to affirm the 'purposeful design', " hence such claims of 'random chance' make no sense. There must be a sophisticated source behind the creation, irrespective of our lack of ability to quantify it."

It is remarkable to point out that just like the patriarch Abraham, Isaac Newton held the view that the Creator is an indivisible Unitarian 'purposeful power' thus against any plurality of any version such as the controversial 'Trinity' within this 'purposeful design'.

The 'purposeful design' of the universe provoked a serious intellectual deduction to conclude that the universe, with its planet Earth in particular, must be too specific and calculative to be simply accidental. Subsequently, we always wonder if the whole purpose of our existence is to engage and audit the following puzzle: if there is 'God' then is he asking us, 'let us see if the humans endowed with high intelligence can recognize that the universe particularly the planet earth is too good to be 'random?'

There is no shortage of scientists who concluded the universe with its planet Earth is too accommodative and supportive for life to emerge and perpetuate to be random, which means 'unintentional or chance', etcetera. Physicist Brandon Carter's 'Anthropic Principle', the Strong Anthropic Principle (SAP), has engaged with this 'intellectual premise' and subsequently rendered the planet

Earth, and the way it supports life cannot emerge by chance. It must have been purposefully designed in order for life to come into existence within it.

Further, this is, to date, the most potent realization due to the fact that the current telescopic advancement and its capture of celestial bodies surpassing two trillion planets in the observable universe because the universe is too large for our telescopes to capture entirely compounded by its further continuous expansion, there is no substantial planet where life is flourishing just as planet earth.

Since we as 'humans' cannot create a planet of 'any sort', had there been only the planet Earth in existence with the absence of any comparative mechanism such as any other planet, it could have been closer to remotely entertaining the 'random theory'.

However, with existence of other planets, for instance, in our solar system, precisely proves to us 'ruling out' any other possibilities. Mercury being the closest to the sun, life as we know it cannot come into existence. Just its climate alone is sufficient for this demonstration, reaching during the daylight 430 degrees Celsius (800 F) versus during night the cold temperatures surpassing minus 180 degrees Celsius (minus 290 F) due to its disability of heat atmospheric retaining capacity.

Let alone the farthest planets, such as Neptune and Pluto, respectively, with an average temperature of minus 214 degrees Celsius (minus 354 F). Neptune maintains an average temperature of minus 232 degrees Celsius (minus 387 F). The planets closest to planet Earth's proximity, such as Venus and Mars, are showing us life cannot be supported as well within their temperature and environment.

Mars's average 'median' surface temperature is negative 65 degrees Celsius (minus 225 F), while Venus's average temperature is scorching 464 degrees Celsius (867 F). It is hotter than Mercury because its atmospheric design is thick and made out of greenhouse gasses such as carbon dioxide in conjunction with nitrogen and sulphuric acid. If you ever wonder what the fate of a planet might be like with Earth's rising greenhouse gases. Just draw an inferable conclusion from the numbers of Venus. However, the point is clear: the 'purposeful design' leaves us with a head-scratching question of how this can ever be 'a random chance'.

Further, the cluster of an uncountable number of planets and stars out there with no verifiable life anywhere clearly reinforces the notion that planet Earth did not come into existence by fluke so as ourselves as its inhabitants. This means the homework had been prepared for us just like the provisions of food we eat all we need to do is 'cultivate and cook' to reap the benefit.

The planet Earth's distance from the sun cannot come in any other way but 'purposeful and thoughtful' placement, irrespective of whether we know who is the cause behind it in the 'ordinary' meaning of the word, such as the way we recognize and know our parents and so on, it cannot be a chance!

If we give attentive contemplation in the simplest but most effective manner of what constitutes 'habitable planet Earth' for humans to be able to sustain life and the requirement of their culmination is not only a thing to marvel at on its own, but simply it provokes our curious minds to infer its factual fabrics.

Consequently, for life to be possible, the planet Earth combines the correct distance from the Sun. It has the protective ceiling of a magnetic field to shield the harmful solar radiations and elements, just like in our vernacular; we place a roof over our heads to protect us from the elements of the weather.

To say the roof of the planet 'the magnetic field' is a random chance is to say that my accommodation's roof came into being by a random chance. Denying your effort of renting or buying your property, further concealing the purposeful design by the developer of that particular property overarching the engineering, architect, the excavation, laying the foundation, framing, roofing, wiring, not to mention planning permissions and other formalities.

The planet Earth is developed with atmospheric insulation that keeps the weather in check so that we can live within it. If you were to entertain the random chance, it is as though we make the presumption to say people who are wearing coats to warm themselves in the winter is also a random chance, precluding the effort of buying that coat for that purpose. Let alone the manufacturer which produced it with the intent to supply it to you and me for the same purpose.

The planet Earth displays the precise chemical ingredients for life to potentially not only emerges but to thrive. Such ingredients prerequisite to life include the abundance of water and carbon without which you and I may not be here. Also, the planet Earth's purposeful multifarious cycling processes, which produce the minerals and energy essential to life to blossom, including the nutrient cycle such as the carbon cycle, the rock cycle, and the water cycle together with their dynamic interactions and many more cannot be the work of 'aimless happening'. The random chance's proposal in order to dupe the human mind with such iteration ought to be fruitless for the thinking immortal intelligence.

The immortal intelligence is the concept that never surrenders itself and, therefore, never surrenders its intelligence without rationalization and reason to any premise of doctrine. As a result, it inquires proofs in conjunction with logic in order to reach such conclusions.

Therefore, ask yourself which 'random chance' brought for you your property, your car, pays consistent income into your account, and most of all, which 'random chance' ever brought into existence an ever living human being or pardoned the concept of 'death' for any person in history?

If the answer is none, then the ultimate veracity of random chance must be a myth clothed in fancy dress; it is an unsubstantiated 'elegant' fairy tale which asks us the blindfold surrendering of our immortal intelligence.

Without a doubt, the thinking minds of mankind, whether philosophers of the past or contemporary scientists of today, whether the rich or the average ordinary people have been inquiring about this question; some called it 'rightly' the big question, others the purpose of life or the point of existence thus this created over the years a dichotomy of two camps 'random chance versus purposeful design', chaos versus order. The 'Random chance' camp is those who argue there is no point to creation, everything has randomly appeared via accidental chance or sheer luck and ultimately will disappear without any purpose.

This is contrasted with the opposite camp, which is those who emphasize that the creation of planet Earth and its inhabitants are designed and executed too purposefully beyond all reasonable doubt to be mistaken with any other conceiving possibility other than 'a purposeful design'.

However, this is not to reignite the two strands of the past, the 'creationists' versus the 'evolutionists', although some comments may overlap due to the common topic at hand. As much as the artificial orbit would like us to see it 'as it were' religion versus science, the immortal intelligence takes the analytical view of non-automatic endorsement to both camps: as most religious premises purely rest their justification on 'faith' alone should fail as much as the evolutionism premise which intends to rely on a blanket application to everything in existence have evolved based on random chance with threshold limiting its scope 'effects' only thereby omitting the trigger cause behind the effect altogether, rather the immortal intelligence aims to engage with the origin, the source, the thing that causes the evolution itself where evolution took place because not everything is substantiated to be directly due to evolution from another 'state'. In other words, the aim of immortal intelligence is to engage with what triggered the beginning, the ground zero of existence. For that we will apply our immortal intelligence to the analysis of 'the Big Bang' theory and what it can tell us about our origin.

Further, immortal intelligence distinguishes between a conclusive science from an inconclusive hypothesis conducted in the name of science because ordinary people such as you and me who are not competent in the subject matter, which means those who are not scientists themselves may give uniform regard between a conclusive science and conjecture of people who use the name of 'science' to sell their opinion as a compelling directive.

There is 'almost' authoritative rhetoric behind the word science even when it is speculative research because it occurs in the scientific investigational realm and carries the domain name of science. Consequently, the common people are intimidated by it. Science itself is evolutionary as the evidence becomes clearer through new technological, intellectual and scientific discoveries.

A conclusive science can neither be complimentary nor synonymous with the irrational of 'Atheism' since atheism by definition is framed as the certainty there is no God, nor there is a source of creation. It asserts with clarity there is no foundational source 'God' which the universe may originate form. Such a premise of thought cannot justify its claim, simply because it rests on fundamentally unsubstantiated, frankly, a flawed reasoning. The concept, in and of itself, was built on bogus guesswork and conjecture or at least inaccurate, imprudent and not well thought out grounds of logic.

In order a reasonable person to conclude the absolute absence of an object of interest (OI) 'whom' we are investigating such as the task at hand whether there is God or not , we must be able to search the entire parameters of existence, in this case 'the entire universe' . Without complete access, a claim of total non-existence exceeds the limits of rational inquiry.

An example is provided: Many of us have played hide-and-seek with siblings or friends. Imagine a parent and child playing in a two-floor house. Suppose the child is too young or physically unable to climb the stairs. The child searches thoroughly on the ground floor, examining every accessible space. Yet, despite this effort, it would be entirely unreasonable for the child to conclude that the parent is not in the house. Why? Because there remain entire floors upstairs beyond the child's reach. Now imagine further that the hidden parent continues constructing additional floors faster than the child can grow strong enough to access them. Or suppose the parent purchases neighboring land and builds another house altogether. In such circumstances, it becomes impossible for the child to rule out the parent's absence with certainty. The limitation lies not in effort, but in scope, capability and access.

Similarly, the universe is expanding at a rate that far exceeds ordinary human comprehension. According to measurements of the Hubble constant, the expansion rate is approximately 70 km/s per megaparsec, with one megaparsec equaling about 3.26 million light-years. The observable universe extends to a radius of roughly 46.5 billion light-years. Due to this unprecedented cosmic expansion which can increase even faster in the future, parts of the universe recede from us at an effective rate exceeding the speed of light, in our layperson's terms, the size and interior space of our universe is expanding over three times faster than the speed of light which makes the current science incapable of disproving the existence of God. This is precisely why scientists employ terminologies such as "observable" or "known" universe, because let's face the truth; there remains an immense domain beyond our observational reach. Quite simply, there is far more that we do not know than what we do know!

The current rate of expansion places certainty beyond our present-day scientific and technological capacity to fully comprehend, let alone exhaustively investigate. If we cannot access the entirety of existence within our universe, how can we conclusively rule out the existence of God? Under such limitations, atheism, which is argued for, the scientific disproof of God's existence, fails to justify its claimed objective.

With this clarity before us, how can any reasonable person definitively rule out the existence of God? It is proven impossible, much like the child in the earlier example. This difficulty is compounded further by the possibility of a multivers, the theoretical suggestion that other universes may exist beyond our own. If that were the case, the scope of reality would extend even further, just as the hidden parent could build another house on neighboring land beyond the child's awareness. In such a scenario, the claim of total absence of God along with atheism dies; it becomes absolutely unjustifiable, un-defendable and nonsensical.

Any scientific position held on the basis of a hypothesis alone would fail. An example is given: the scientific community held the view that the sun is the centre of our universe and the planets supposedly orbit around it, a phenomenon known as 'Heliocentric'. However, it was not until the advent of new technology such as the 'telescope' that showed the more accurate reality of the issue that the sun is not only fixed but not the centre of the solar system or the 'universe' as it were at the time.

Consequently, through the observation of a telescope, our solar system is 'suddenly' a tiny particle in the observable universe; an area known as 'the barycenter' is the centre of our solar system. The point is science can be an ally to the application of our immortal intelligence only when it is conclusive and accurate rather than a plain hypothesis.

The Immortal Intelligence Engages with the Big Bang Theory

This is not a teaching summary of what the Big Bang theory is, nor should it be, as there are plentiful places for that if you are interested, but rather this is to see how the Big Bang theory or the science of Astronomy for that matter can shape our understanding towards the source behind the creation.

Not only to appreciate how magnificent and massive celestial concepts are but to make sense of whether the calculated order of the observable universe rests on one power or multiple different sources. It is therefore relevant to learn together the concept of 'respective reality'; that is, you will focus on the reality as it concerns you or in so far as the scope of the core object you are inquiring upon.

Hence, being fully cognizant of the entire field of research may not be needed for your concerned goal. Therefore, it will not be a condition to form our view. A demonstration is provided: let us say you are a high school teacher who intends to receive a pay raise; learning successful strategies which generated more wealth for a self-employed cook who is on call for catering such as events and weddings is not your 'respective reality'.

Although it may be interesting, however, the strategy to widen the volume of customers and clients for the cook may not be applicable to your sector, line of work, your 'respective reality', henceforth ought not to be your concerned respective reality. Additionally, to investigate strategies utilized by everyone who wanted a pay raise may be 'unworkable', untenable, and simply require more resources than you are capable of allocating at that given time.

Adopting our 'respective reality' in the subject matter, we will engage with the Big Bang theory and Astronomy at large in order to form a sufficient foundation to make a coherent sense of whether there is authority behind the creation. Whether this authority is a singular source or multiple sources via deducing, inferring and utilizing the information modern technology revealed to us. This means our 'respective reality' does not need to encounter either every celestial object in the universe nor everything about the unseen existence of 'dark' matter and energy behind it, etcetera.

The fact that the universe is expanding excessive distances per second, considering the Hubble constant measurement, which yielded the rate at which the observable universe is expanding and later corroborated with NASA and the European Space Agency (ESA).

The known universe is expanding 46 miles per megaparsec, which is equivalent to 1. 4 billion light-years. Just to draw a mental picture of the sheer magnitude of this speed, one light-year is 6 trillion miles in distance. This is coupled with the new celestial bodies coming into existence every time, which is a gigantic amount. The Milky Way Galaxy alone is believed to have seven new stars being born every second, which is the reasoning behind why we should base our analysis toward our concern, hence our respective reality.

The Big Bang theory points to a colossal explosion taking place approximately 13. 8 billion years ago, while all that existed were gases only. Although the term Big Bang theory calls itself the occurrence of rapid expansion at once, which should be a more accurate explanation, the fact that highly dense and heated gases erupted at once and then, if it looks like it, feels like it and acts like it, then we will call it an explosion.

However, the problem is this is not an explosion as we know it, like a bomb went off, suggesting that its characteristics are uncontrollable. It generates chaos with the ultimate result of destruction, which would aid the concept of random chance—it is a coincidental eruption, luck, fluke, and the like. However, conversely, the Big Bang eruption is quite the opposite of the above.

Despite its heated gases 'explosively' erupting with great force, the net result is building, producing, manufacturing, generating, forming, steering, molding, and creating stars, planets, solar systems, etcetera. This, therefore, is a rather purposeful explosion, which may explain why this explosion has a source behind it 'however unknown' since it has a goal of becoming something, and the paradoxical relationship it has with the literal meaning of an explosion, like a gas tank exploding, never creates a new meaningful thing.

The above has similarities with the purposeful combustion engine explosion, which creates the energy needed for the motion of our everyday cars. The car engine is no different than an exploding bottle. The gasoline enters into the engine and mixes with oxygen. The spark generated by the spark plug instantly heats the hybrid, therefore inflating gases and causing an explosion. It is this explosion that forces the pistons to move fast, as they are connected to the wheels of the car.

This, in turn, is what moves our everyday cars into the motion of forward and reverse as we desire. As a result, realizing there is an internal explosion transpiring every time you see a gasoline car on the road never makes you think that the car is moving via 'random chance' or a fluke because it is a purposeful design altogether, including its explosion, just like the Big Bang is.

Subsequently, the immortal intelligence requires us to draw a parallel of the two and conclude the Big Bang has a cause behind it, just as the combustion engines of our everyday cars, even though we were not an eyewitness to see the engineer who built the car.

However, by virtue of its purpose, we can safely conclude that there is a cause and effect here, the cause being unseen to us, the engineer who built the engine. Hence, the effect is the result of a purposeful engine. Then, if the planet Earth is purposeful in design, just like the car engine, and the Big Bang is the method in which it is heated and expanded, it makes it an obvious framework that there is a greater unseen force working behind the curtains causing it.

Why do we not see this force behind the curtain? The simple answer is this force designed itself to be hidden, otherwise, we would have seen it as we see stars like the sun and moon without any interference. The witnessing or comprehending of the creator behind the Big Bang. We will engage with this concept later as we progress.

Another scenario to draw a comprehension on the subject is the situation where a firefighter investigates the typology of fire, whether it is incidental or arson; fire 'in and of itself' is a destroyer and can be utilized for such a purpose to destroy. However, the authorities do investigate when a fire is suspected to be arson and determine without seeing the causer behind the fire whether this fire is arson or not, which broadly takes the meaning of intentionally starting a destructive fire.

The underlying cause is the 'causer' intended to ignite the fire, even when it destroys. As the arson investigator carefully scans through clues which show the nature of the fire, whether any accelerants were used or not, but most of all, the investigator carefully looks into whether there was a purpose to this fire or if it was a random chance. Was there a burn pattern organized, with starting and ending points, which usually carries a purpose to destroy evidence or property?

The Big Bang, therefore, is a purposeful act which generated our intricate planet Earth with its unique properties necessary for the existence and sustenance of life. These are literally 'but a few' methodologies with which to examine whether there is 'a causer, ' a causative entity behind the Big Bang due to its purposeful behaviour and characteristics regardless of its starting nature of what appears to us as chaotic 'heat, gas, and explosions. '

Now that we have established that there is a causer behind the emergence of the universe even though this causer is unseen, with that recognition, we can also deduce to rule out that we as humans were any part of this early construction. Hence, this entity is not, nor can it be, a human being. The next natural question the immortal intelligence aims to ask is to determine whether there is one almighty entity or maybe several different powers who have concocted harmonious systems and grids to produce a purposeful universe, including planet Earth.

Before we engage with this fascinating second stage of our inquisitive probing, it is time to introduce the conceptual methodology of 'impact recognition'. This is the methodology required to determine when the cause behind the effect is invisible and unquantifiable to the literal vision, such as the case at hand.

This methodology takes away the requirement of literal viewing as a precondition to forming mental recognition of an object. That is, you should be able to recognize an object via its impact rather than its image. Despite this matter being discussed in further detail later in our discussion, a foundational iteration is given: Although it is not precisely synonymous like for like, nonetheless, it provides conceptual appreciation.

The vision of the blind person leading life is not based on forming recognition on the basis of image but rather calling to task other faculties of his/her intelligence, including but not limited to sound, touch, smell, and most of all, the intellectual ability to calculate, deduce and classify any object the blind person encounters with.

Yet still, many people with this disability are advanced and accomplished in their own right without ever resorting to the requirement of an image for their recognition to acknowledge any objects.

There are abundant examples out there, these include from the politician David Blunkett whose duty was to safeguard the British people as the Home Secretary despite he was born blind, and the famous singer Stevie Wonder, who was also born blind, which entails they never required a conceptual image to process the reality of existence, to the lesser-known Marla Runyan, three-time national champion of marathon running who achieved this while legally blind.

Other significant people with vision impairment include the great artist Claude Monet, who continued to paint even after becoming sightless, and The American President Franklin Roosevelt (FDR), who progressively suffered from impaired vision. Please note that the latter two were visionary impaired or became blind at a later stage. The list is inexhaustible; there are many more people with sight impairment who have been scholarly and well-endowed in history. Unfortunately, it is hitherto beyond the scope of our discussion to enlist them.

Let us pay close attention to our solar system, which encompasses the well-established eight planets and their star, the Sun. Although NASA has trajected, the number of planets is more than previously held. Each of these bodies travels on a preset path similar to the roads which we drive on in any of our major city grids. They circumambulate or make a circle in their orbit, much like our roundabouts in our city driving scenarios.

Both our roundabouts and the natural orbit have been designed with the same purpose, which is to define a direction of movement so that there is no clash or accident in their respective objects, cars on the part of our streets, planets and a star on the part of our solar system.

Our street grid systems, including their roundabouts, likely have many different authors despite their consensus toward an engineering design 'blueprint' and rules in relation to the framework of the road. Further, with certainty however, each car on the road is controlled by different driver 'source' which is the main evidence of cause behind road accidents and collisions.

Despite drivers being trained and licensed to follow the direction of motion on the framework of the road, nevertheless, accident reports are occurring consistently on a daily basis. The accident reports of the US jurisdiction alone demonstrate this fact vividly. Major road collisions, which result in fatalities on average, are numbering between forty to fifty thousand a year. The general collision numbers which do not cause fatalities are gigantically higher. Conversely, we cannot say the same about the planets and the Sun making their motion on their designed framework known as 'natural' orbit.

The immortal intelligence acknowledges if there had been different drivers/sources who are controlling and navigating the different planets and stars in our solar system, such as the planets, the Sun and the Moon, there would have been some frictions, accidents or clashes or gridlock traffic jams due to the diverse minds attempting to converge on the framework regardless of how systemic the road framework is. An example is granted: just like our road cars, if the planet Earth is controlled and navigated by a different entity than planet Mars or any of the other planets for that matter, such as Venus, Mercury and the like or the asteroid belt with its buffer zone function would have collided or clashed with each other with similar outcomes just as our cars with different licensed drivers conducting their systemic motion on road grid system built for that purpose.

The fact that there is no discrepancy in the way these planets observe safety when they are travelling on their equivalent road system grid known as the orbit leaves no other explanation but to conclude the entity controlling the system must be the same 'mind' from the same source due to the level of harmony discharged and observed. As such there cannot be plurality with the source of control behind the safe operation of our solar system.

Yes, even the best of friends, allies, partners, spouses, and colleagues cannot maintain a 'dissent-free' relationship. The kind of absolute harmonistic behaviour in which these celestial bodies are driven in their paths is not only a testament to their predication on the same source but this entity controlling and navigating them beyond the human eyesight is brilliant in maintaining this order consistently for millions of years without any discrepancy let alone its preventive measures of a catastrophic clash, which would result in an existential destruction to planet Earth and its inhabitants. Come to think of it, we ought to be thankful that it is rather one entity behind the power rather than many, which would inevitably be susceptible to disagreement or conflict.

Just take a look at our democratic parliaments: from time to time, we see polarized disagreements along party lines or political views that our parliaments become 'dysfunctional' momentarily as we call it 'hung' parliament or the US Congress terms it 'filibuster', it is not unheard of or unprecedented, to see in some cases, the most supreme branch of a democratic governance system the Parliamentarians or Congress persons depending on which jurisdiction you reside, coming to a physical confrontation to demonstrate their discordancy.

Parliamentary brawls just happened in Italy in June 2024, so as in Canada in December 2012, in the EU in 1988, Japan in

September 2015, South Korea in 1966, and in the US at the Federal or State level no less than twelve times the last of them being January 2023 at the House Speaker election, which is irreconcilable with the 'singular command power' behind the smooth running of our solar system.

A parliament, in substantive democracy, represents a 'plural command system' where a group of elected people representing different sects of the society set policies which ultimately become statutory legislations on a consensus and majority basis, which is a stark contrast to the 'singular command power' which entails there is no room for difference of opinion and views on the subject matter, a system so directly fitting to our solar system as 'glove and hand'.

Had there been a plural authority like a parliament presiding over the governing of our solar system, we would be in a colossal turmoil beyond measurable proportions when and if they reach their equivalence of 'hung' or brawling parliament, much less their equivalent of World Wars Ones and Twos, etcetera.

As if the above reasoning does not suffice our quest, now we will pay close attention to the Big Bang event itself in order to determine what clues it can provide to us about the subject matter. The unseen power behind the creation provides us with compelling clues in order for us to reach an informed and conclusive determination as to the prospect of our beginning.

That is, the initiation of our universe, an illustration is given: have you ever wondered why movie-playing instruments have been designed with the feature to rewind the movie? The purposeful designer behind such instruments, just like the universe, intended to confer upon the viewer the power and ability to trace back the storyline in order to make sense of the point of origin which the author of the movie intends to deliver to the audience.

In data interpretation, the immortal intelligence differentiates between collecting data from its interpretation; therefore, any reasonable human being is equipped with not only the fundamental intelligence to apply any data according to his/her deliberation but also possesses the right to uphold such autonomy. Yes, our math teachers can educate us in arithmetic calculus and how to carry out numeric calculations such as addition or subtraction, but the right to apply these numeric functions to the way in which we would like to spend and allocate our money and resources has always been lying and remaining with us, in that, we always have been intelligent autonomous beings.

In other words, it has always been up to us to decide and prioritize the allocation of our resources 'money' at any given time and scenario based on what we deem important in those specific circumstances. For instance, we decide whether we should go on vacation this summer or save the money for something far more important, perhaps to buy our first property. We do not rely on our math teacher who taught us the foundations of calculations to make such impacting choices for our lives.

Henceforth, the scientific data gathered from our telescopes would give us the foundational information and facts, but the ultimate decision to apply them ought to be within our intellectual deductions and reasoning even when we accept the hypothesis of others, including my own. That is, if I am not making any sense not only do you have the right to disregard any of my proposals but the right to scholarly critique them too.

As therefore demonstrated above with the illustration of movie-playing instruments, whether your Netflix or old-school cassette players, they always have enabled us to rewind the sequence of events so that we, as the viewers of the movie, can trace back in order to authenticate the cause and reasoning behind the latter incidents, events, circumstances and situations etcetera.

Similarly, the universe empowers us to trace back and detect the sequence, triggers and causes behind the making and shaping of our universe as we know it today, for instance, how old is our universe, what took place at its starting point, not only that but is there death taking place to these massive celestial bodies up there too such as the phenomenon of 'Supernova' explosions which amounts to be the last 'death rattle' emanating from the dying creatures at their final moments on planet Earth pertinent to humans and animals alike except the privileged few who are reprieved.

With our contemporary scientific advancements in Astronomy, such as the conjoined initiative of NASA's Goddard Spaceflight Center and Princeton University that led to the emergence of the

Wilkinson Microwave Anisotropy Probe (WMAP) satellite as the astrophysics community call it 'the cosmology precision era' which signaled a new age of less reliance on hypothesis and theories as the precision calculations of distance and time reached at all-time high within range of plus or minus few percentages.

This further strengthened the ground-breaking Einstein's theory of general relativity, which suggested by calculating the degree the universe's 'matter and energy' accelerated to expand in the past, we can reverse the clock back until we reach the time the universe was 'zero' size just as afore pointed out we can press the 'rewind button' of our electronic device until we gain knowledge of the genesis 'the origin', the beginning of the universe when there was nothing.

The WMAP satellite took the matter of accurate computation between 1 to 3%, which is the closest we have ever been to certainty. Consequently, as WMAP measured the composition of the universe and its expansion rate, that is, the composition of matter and energy density and then reversed back the time utilizing Einstein's general relativity, you would reach the Big Bang's point of singularity with an accuracy of 0. 4%.

In other words, as we trace back the expansion of the universe just as we press rewind on our movie player, we can obtain the most vital clue of the source, the cause behind the Big Bang and the birth of the universe, which takes us to the point *'Singularity'* a unique word, a significant description, more importantly a unique meaning, which entails from a single outlet appeared a single dense and hot point which rapidly created space with it came matter, gas and energy to evolve forming the universe as we know it today.

This affirms that everything in existence in our universe, whether solid, energy/gas or liquid, must go back to this 'singular' point. Therefore, its source of beginning zeroes in on the fact that it cannot be added or subtracted hence remains a 'singular source' from which everything in existence is derived without any reasonable doubt at all. Astonishingly, the same word is given for the explanation of the origin of the universe and everything in existence in the Koran.

It tells its reader the source of creation in the heavens and the earth, as well as everything in between them, goes back to singularity 'Ahad' as it calls in its Semitic Arabic language of origin. Say your lord is 'Singular', it says to the reader in verse one, chapter 112. The word Lord in the Semitic languages refers to the source for provisions, which applies accurately to the 'point of singularity', meaning from this point came all provisions, whether solid, energy and gas or liquid.

This Koran 'book' confirms books that came before it as it proclaims in verse 3 of chapter 3, such as the book given to Jesus Christ as Jesus says in Mark 12:29, which repeats Deuteronomy 6:4 our lord is the singularity 'Echad' reaffirming the same meaning. So, as Biblical Moses in the Torah and patriarch Abraham who voiced an identical description of the beginning of the universe being 'Singular', our lord is 'Echad' in Deuteronomy chapter 6, verse 4.

These sister words of 'Ahad' and 'Echad' in the Semitic languages appearing in the Koran and Torah, respectively, for the description of the beginning of our universe, equates with the fact that everything returns to this Singular point, which means everything is from a Singular Source, derived from this Singular foundation.

Therefore, as such, whatever can lay claim to having originated the beginning of our universe, 'God' or any superlative adjective we may tribute to due to its unique capabilities such as limitless power, purposeful designer, intelligent designer and all-powerful source, etcetera must match and correspond to this underpinning.

This underpinning consequently precludes things and beings which came as the result of the Big Bang, such as the celestial bodies who are facing death themselves in the form 'supernova' or black hole. It also must eliminate any human God, 'no thanks' to disappoint many, but the fact is human species, let alone attaining a claim to be a candidate of the original source, were not even in the realm of existence at the early stages of our universe.

The same goes for everything which intellectually amounts to being 'less' than human beings in their natural composition, such as other animals, plants, etcetera, as they were nonexistent even after the Big Bang when our universe was plasma, which is the composition of electrons and nuclei, in other words, all the naked eye could interpret at that time was 'thick smoke' after the Big Bang event, and that was all that existed in our universe.

Astoundingly, the traces of this concept of a 'Singular source' of supreme power of origin are found even in the most popular 'Hinduism' books such as Bhagavad Gita chapter 10, verse 8. Despite murky surroundings of plurality and multiplicity of deities proliferating down the line surpassing in some respects within this philosophy three hundred thousand nominal gods. It says to the reader to attain knowledge of the 'singular' supreme lord of the worlds in Bhagavad Gita chapter 10, verse 8.

When the reality is such that one has to admit it, for instance, 'the time' we are in at any given moment during the day, it is either night or day to you. If not, it must be the brief sunset or sunrise duration, there is nothing else that can exist besides these mentioned above scenarios.

Similarly, there are situations and circumstances in which one can reach a 'decisive' conclusion which does not allow any other probabilities or interpretations that can lead to different positions on the subject matter.

People have been divided historically into one of these two camps: those who are explicit affirmers of the fact of a 'Singular Source' of power irrespective of what they called it. Such people comprise intellectuals of the highest human achievements such as Muhammad in Islam, Jesus Christ in the Gospels, Moses in the Torah, as well as the patriarch Abraham. Ironically, the man Siddharth, otherwise known as 'Gautama Buddha', was silent about the subject matter. As such, it is safe to say that there is no record of 'himself' engaging with the question of how did it all began.

The explicit affirmers also include great scientific minds who contributed to our human advancements, such as Isaac Newton, Galileo Galilei and Albert Einstein, just to name a few. Additionally, there are fairly contemporary scientists who were explicit affirmers, as well Professor Christian Anfisen, 1972 Nobel Prize winner, who said, "We must admit that there exists an incomprehensible power or force with limitless foresight and knowledge that started the whole universe going in the first place".

Furthermore, the renowned MIT physicist Professor Emeritus Ulrich Becker, a Nobel Prize winner in 1976 for the discovery of the J particle in high energy physics, said, "How can I exist without a creator? I am not aware of any answer ever given". Prior to them, there were other great scientific minds who contributed significantly to human development, such as the father of 'algebra' and 'algorithm' Muhammad Al Khwarizmi. These great minds were all explicit affirmers of the 'Singular Source' of power at the starting point of our universe.

The second group are the implicit affirmers of this irrefutable reality irrespective of whatever position they may hold on the question of the supreme power, nature, etcetera or any other descriptive adjective conferred on this phenomenon. Virtually no scientist can reasonably refute the Big Bang and, therefore, its origin of 'Singularity', including those who claim to be 'atheists' because the ultimate truth of our point of origin is such that you are compelled to conform to its 'realism'. Which is why Steven Hawking admitted and conformed to this reality implicitly despite his intention was to repeal it as an open atheist in Singularity Theorems and brief history in time,

If this is your moment of realization, if the light went on in your head at this junction, then congratulate yourself you have taken your first of the two steps required to activate your 'immortal intelligence'. Which means you came to the conclusion that there is an 'unseen' or 'imageless', at least in so far as we can visualize according to our capability with regards to the source of power behind this purposeful existence we call the Universe, particularly pursuant to our solar system and the planet Earth which is uniquely created and placed so that life can thrive and be supported.

Consequently, this calls for two questions: for what point this purposeful design is supposed to serve and who is 'correct' amongst those who are claiming that this purpose is within their philosophy, religion or organized thought out there? Therefore, we will briefly authenticate or rule out with evidence beyond 'blind faith' or 'blind hypothesis' regardless of in whose name it is claimed, even if it is in the name of science. If something requires a blind following, a pure 'leap of faith' for you to agree against reason, logic, and sense is just a tough bill to swallow for the 'immortal intelligent mind'.

The Immortal Intelligence Engages with Neutralism.

Neutralists rely on the claim that there is no empirical research to prove whether there is a 'visual' creator behind our existence or the emergence of our universe; however, on the other hand, they dismiss the atheistic view that all this delicate or purposeful creation appeared via 'a random chance'—an area we extensively discussed earlier. Therefore, in many ways, they justify their intellectual curiosity in the subject matter, claiming that their position to remain neutral is the 'safest bet' and consequently perpetuate their lives, never to engage with it until they meet their end.

Having said that, the downturn of this mindset is that they develop a culture of 'reactionism' as opposed to becoming activists themselves. They wait for others to engage with this all-important topic and, in turn, spoon-feed them the answers. The Immortal Intelligence, however, takes into account two factors that make 'Neutralism' unpragmatic. As such, the Immortal Intelligence factors in the motion of time and finality, which are not stationary at all, and their impacts possess an absolute power over all humans without exclusions whatsoever.

In terms of time, it is unanimously agreed that it is the one thing we do not have, and that is why almost all of our technological advances were purported to elongate time in order to make time more abundant. Airplanes reduced dramatically the time needed to travel in the past when the mode of travel was confined to horses/camels and boats/ships, etcetera.

Similarly, electronic mail 'emails' shortened drastically the time needed to deliver a letter between people and organizations. The list goes on and on, including cars, telephones, mobile/cellular phones, etcetera. Despite all these achievements to make time abundant, like a buffet table in a restaurant, it is proven to be the opposite in terms of its impact on people. Time has increasingly become preciously scarce in our modern and techie lives.

In terms of finality, our end, which is commonly known as 'death', is the most individualistic exit, the utmost sole project you will ever be assigned to without any choice to opt-out or even to delay it. Due to its individualistic nature, you cannot delegate it to anybody else in playing the waiting game. This means no one will carry out your awaiting task of death for you—not your religious clerk, not your atheist scientist, not your golf buddy, not even your spouse or parents. Therefore, the Immortal Intelligence remarks that you cannot afford to wait for someone else to answer this question. As such, you must become an activist and seek your own individual answer to this monumental intellectual consideration.

There is the proposal that our existence serves the purpose of seeking an answer to this big question. As such, therefore, it is an experimental test to see whether the most intelligent beings on the planet Earth—the humans—can find a reasonable answer to this epic intellectual curiosity.

An example is provided: imagine you are sitting in the middle of your university exam and refuse to answer any of the questions, justifying your un-involvement through 'I am awaiting someone else to answer my questions. ' Even if your student peers answered this question, they did so for their exam without any concerns about you because their immediate and most important focus and task is to see themselves pass the exam—not to find for you answers that you 'yourself' neglected to seek for with your own effort.

Another consideration is conferred: just imagine you are born inside a compound that has parameters which cannot be passed because life would cease to exist. The compound accommodates all needs and requirements of your life to exist, such as enough space to move around and exist.

However, when it comes to your provisions for your livelihood 'sustenance' , you have learned of a drop box where you collect all of your necessities for your life support, and you collect it just like your Amazon delivery person left your items for you, or the post person dropped your letter in your enclosed home. You would definitely possess a strong longing to meet this delivery person, perhaps to acknowledge and thank 'him' or even some of us would like to complain to him due to some disappointments along the way.

Despite everything of provisions being a donation in the first place, such as the rain that gives us the vital water we need for our existence, the crops that give us vital food for our existence, the nutritious fruits and vegetables that benefit us in all sorts of ways from bolstering our immune system to keeping us healthy and enthusiastic, and the essential proteins and amino acids from meat and dairy products as well as plants.

An immortal intelligent person would never hold the belief that 'for one minute' you would get orange fruit during the winter season with its high vitamin C content of infection-fighting properties, amounting to one medium orange providing 83 mg of 92% of your daily value (DV). Oranges, along with other citrus contents, are designed to fight winter flu viruses via increasing white blood cells known as lymphocytes and phagocytes, which in turn help protect the body against infections during the flu season.

Contrast with watermelon in the summer season for its high contents of dehydration suppression as it carries 92% of the water content, which contains vitamins A and B6, amongst others. Further, it contains an organic pigment called lycopene that shields people from sunburn due to the heat of the sun during the summer months.

Any thinking intelligent person would never conclude the above scenario is just random chance without 'purposeful design'. Metaphorically, these are some of the beneficial food stipends we are receiving from our drop box without seeing who it is that is giving these provisions in different seasons of the year.

This is the concept the Immortal Intelligence would call 'impactism, ' which denotes the ability to identify an object with its impact rather than its physical existence, which will be expounded on later for further engagement. Now imagine the compound is the planet Earth in which we live.

Therefore, Immortal Intelligence finds this strand of 'Neutralism' to be lethargic, lazy, unwarranted, and most of all incompatible with itself. Conversely, the Immortal Intelligence encourages people to be active in searching and engaging with this big question, even if you think it may not yield any definitive answers to your curiosity.

Those are the negative thoughts you need to overcome for your greatest achievement ever. Yes, figuring out your purpose of existence is the greatest achievement of all time. Words cannot express the magnitude of this goal. It is greater than any award 'the artificial orbit' wants to inundate you with.

It is greater than the Grammy Awards, the Medal of Freedom, the GBE of the UK, it is greater than any sports championship you may be competing in, it is even greater than the Nobel Peace Prize, except where the topic of the award is finding an impartial and truthful answer to this very question. Otherwise, none of the above will go with you to the next journey of your life after death.

The Immortal Intelligence Engages with Agnosticism

The sheer evidence of direct and indirect support to existing purposeful design compels agnostics to the plausible reality that there has to be a creator, but simultaneously, the high number of unsubstantiated claimers through various philosophies out there puts them in the same predicament as atheists. Which is to dismiss the candidacy of these gods to qualify the title of the unseen power behind the emergence of the universe and its 'Singularity'. A topic we discussed in detail above.

These candidates include, without intending to offend any of their followers, as humans, we are entitled to choose with our 'free will' error in place of clarity. These beings include animals, plants, and statues created, curved and sculpted by humans, pictures of dead people and other fellow humans who clearly fall well within the vulnerabilities and weaknesses of being human themselves, such as forgetting, getting sick, being impacted by hunger and consequently eating and drinking, which comes with the requirement to need to take the 'call of nature', not knowing the future, and themselves crying out to another deity other than themselves.

All of these shortcomings propel the agnostic's intellectual reasoning to disqualify these proposals out there. This is compounded by the fact there is no visual identification of the unseen creator of the heavens and the earth.

In light of the above circumstances, the Immortal Intelligence broadly sees this strand as a step in the right direction because, as we described earlier, intelligence is not only positive by gaining benefit, profit, etcetera, but it is also negative by preventing harm, recognizing falsehoods and scams or the ability to differentiate unverified claims from substantive realities.

This realization is clear to them: none of the above candidates are qualified for the title of the creator of the universe by virtue of who they are, including humans. The claim that God became human is irrelevant because God was not human when 'He' created the heavens and the earth. At the starting point of our universe at the 'Big Bang' when all that existed were plasma and gasses, there were no humans then, nor were there animals and plants, let alone statues sculpted and created by humans who did not exist 'themselves' at that time, thus none of these 'deities' is worth our celebration for our existence simply because they had nothing to do with it.

This leaves us with this question: how to identify and comprehend the only prospect we are left with, which is whoever did it is 'unseen', whoever created the universe is non-visual, and whoever is the cause behind the Big Bang is hidden from the human eye 'at least' up until now.

The Immortal Intelligence Recognizes 'Impactism'

We need to give an elaborative explanation and definition to the conceptual meaning of the word 'Impactism' in this context and why it is a very pivotal intellectual tool for finding answers to certain questions, such as the situation at hand of comprehending and making sense so that the factual pieces fit together in order to get an overall coherent picture of who is and more relevantly who is qualified to be the candidate of the creator of the heavens and the earth that is with such physical and intellectual capability to cause the 'Big Bang' and the like.

The entity we are seeking for is unlike and incomparable to anything else we so far can visually identify or have encountered within sight. To have an idea of its colossal magnitude, one of its insignificant creations is the star we call the Sun; its heat power reaches 15 million °C (27 million °F) at its core and 5, 500°C (9932°F) on its surface.

We define 'impactism' to mean the measurement by which something 'unseen' is identified by virtue of its impact rather than its physical appearance. The irony is humanity has been applying a variant of 'impactism' since the dawn of our existence. Ancient cultures and civilizations have developed methodologies with which to track lost livestock or people by trailing their footsteps; others employed 'impactism' by way of hunting down foes and enemies who have fled with the intention to be hidden from their pursuers.

The fact is it was a trading profession, 'tracking', despite being 'a niche skill'. Actually, people were charged for such services in situations which warranted its expenses. For the 'tracker', the underlying 'constant search' remained unchanged, which was searching for something the tracker had not physically seen before.

The caveat, however, is granted that the tracker had a pre-seen image of his target in this situation. An example is provided: Even though the trucker may not have seen the lost person, Mr. 'John Smith', but has a certain picture of human anatomy. Hence, the trucker has certain parameters of the image to target for in the first place. Similarly, despite 'you' not having seen your postal officer delivering your mail in your Dropbox, nevertheless, you obtain a preconceived visual mental picture of the postal deliverer, which is why this is a variant of 'Impactism'. Perhaps we should call it 'partial Impactism' due to the existence of its preconceived image of the target one needs to identify, even though there is a clear absence of visualizing the particular incident and/or object itself.

However, there are scenarios where the preconceived image is less pronounced and more obscured that we humans engage with to identify and recognize those objects based on their impact alone. Ask any 'ichnologist' how do you come to the knowledge of the unseen target you are conceptualizing without ever having seen their image in the first place.

These branches of science, known as 'Ichnology' and 'Palaeontology', are sufficient testimony against the claim that physical image is absolute requirement for ascertaining the existence of something. It is the ichnologists who explained to us not only the existence of something unseen such as 'the Dinosaur' via Tracks and Trackways but ultimately traced back their footprints to reconstruct this 'phenomenal existence' until we ended up with not only full-fledged beings, but dominant beings, in the animal kingdom that roamed around the planet Earth more than two hundred million years ago.

The discovery of dinosaurs is different to that of ancient trackers. Recall the trackers had preconceived images of human anatomy before they traced the lost specific person or animals, etcetera, while ichnologists who investigated and traced back the dinosaurs had nothing with regards to the preconceived appearance of their target of discovery.

As such, therefore, it is not only disingenuous to require the image of something for your admission to their existence, but it amounts to be no less than an oxymoron to affirm the investigative approach of Ichnology for unseen 'worlds' such as dinosaurs in Triassic and Jurassic periods two hundred million years ago while denying the equivalent trailing and investigative approach could be directed to the unseen power behind the creation of our universe via its impact, an immortally intelligent person does not blindfold his/her intellect in such a manner.

The fact is there are even more obscure existences humanity has discovered than dinosaurs and the like without relating to a shred of image whatever. Ask any Neurologist how do you diagnose or recognize the degenerative illness of Parkinson's. It would be self-evident that neurologists employ the approach of 'impactism' which leads to the examination of this illness's impact on its victims that determines its classifications.

Typically, neurologists look at the hallmark impact of this disease on people, such as tremors, bradykinesia, rigidity, and postural instability. Make no mistake, even scans such as DaTscan, which exhibit the dopamine transport levels, accompanied by the lack of striatum support uptake within the brain, are still confined to measuring the impact of something unseen.

It is widely accepted to acknowledge the existence of Parkinson's to be the unseen cause behind its effects via its impact without ever necessitating the visualization of its direct image, then how is it when it comes to the greater unseen cause behind the creation of our purposeful universe is 'lowballed' shortchanged with far more abundant and transparent impacts and effects than Parkinson's ever amounts, an immortally intelligent person cannot be fooled in such a manner.

If the recognized hallmark of the unseen Parkinson's is tremor, bradykinesia, rigidity, and so on, what stops us from recognizing the hallmark of the 'unseen' creator of the universe and everything within it, including inter alia everything which falls into opposite but complementary pairs such as gender pair, male and female which are found not only in the animal kingdom but also in plants, day and night or the sun and the moon are all pairs or everything living needs 'water' which represents the hallmark of this unseen power behind the creation.

Therefore, any entity which falls within the parameters of this pairing system is created and, therefore, not certainly the creator in the sense of being the source behind the universe itself. Omitting the lower usage of the word 'creator' that mankind attributes to himself relative to his/her capacity. As we say, we procreate our children and so on. An example is given: when the car designer Malcolm Sayer created the Jaguar E-type 1961, the most beautiful car ever made, according to Enzo Ferrari.

The usage of the word 'created' is limited to our human capability because we never created any of the raw materials utilized to achieve the final renderings, such as the metal, leather, etc., nor can we ever contemplate the creation of the human intellect which was used for the purpose to bring out this final 'exotic' product. Certainly, the most pursuant pair to the subject matter is the acknowledgement of this pair 'life and death'.

The reality is we recognize them only through their impact, hence 'impactism', which means these two pairs are not visually quantifiable to us. As such, no subjective measurement can be directed upon them, such as black or white, tall or short, lean or bulky. However, we appreciate their universal effects upon creatures without any exceptions. Whether humans, including atheist scientists who ideate 'random chance' or plants and animals alike, there is no exception or exclusion to this rule that every living thing awaits its death.

The trend of recognizing something through its impact without ever seeing its image is shockingly abundant around us every day, it is even in our everyday vernacular 'lingo'. Any time you lose your internet connection, the signal lost has no image but is recognized by its impact on your connectivity, so as your mobile/cellular phone connectivity when it has no reception, the trend is so copious that our human eye always has its limitations to process an image of subatomic particles such as electrons, protons and neutrons as well as electromagnetic spectrum such as infrared radiations, ultraviolet radiations, x-rays and so on. Other invisible phenomena include magnetic fields and radio waves.

No wonder the Higgs Boson 'sub-atomic' particle that was discovered in 2012 by respectively Peter Higgs and François Englert that led to their Nobel Prize in 2013 was code-named the 'God particle', which corroborates the invisible power affecting things 'particles' while transcendent to the human eye are conducting operations in an apparent manner via its impact.

When funding was sought from the UK government in a larger funding request of 10 Billion dollars for CERN's larger Hadron Collider (LHC) project in Switzerland, then the British cabinet minister for Science William Waldegrave promised a ceremonial reward prize for any scientist who can offer a relatable explanation for this invisible particle's existence in layman terms to the general population. Professor David Miller of University College London (UCL) took the challenge with his famous Thatcher analogy.

He utilized what, for 'all intents and purposes', is the 'impactism principle', which is measuring something hidden through its impact rather than its 'image appearance' by measuring the formation of clusters around them. Professor William called it the 'cluster effect' by observing how clusters reacted and formed to the hidden 'Higgs field', which translated to when an ordinary person comes through a crowd of people. The crowd reacted to the presence of the ordinary person in an ordinary way, lagging interest towards the ordinary person. However, conversely, when a famous and powerful person, hence, the analogy of Margret Thatcher, the then-prime minister of the UK at the time, entering a crowd of people, the people react to her presence in a much more gravitating way towards her just like particles cluster more tightly towards the hidden 'Higgs field'.

Furthermore, the admittance of dark matter and energy is yet potent evidence towards the need to recognize the existence of the 'invisible hand' which is setting and administrating the laws of physics hence keeping our universe afloat so that we can exist peacefully.

As the name 'dark' suggests, it is the recognition of something controlling and administering the essentials of our universe and existence without us gaining access to visualize who it is that is doing all this while remaining hidden. No doubt this sight impediment is intentional. Dark energy has mystified scientists while clearly, through 'impactism', it justifies its existence, its impact is far greater.

It is the force which is behind the rapid acceleration of our universe's expansion at an astronomical rate that no one can understand except through its impact. Furthermore between, dark energy and gravity, which are both something beyond the human eye, are holding our existence together.

Consequently, it should not be an incomprehensible phenomenon when 'the God' is described as the unilateral power of an invisible, hidden force behind administrating and controlling our universe and ourselves. The novel beginner of the universe with its 'singularity' that is based on a single event for its inception and what sprung thereafter to our current status quo.

Now that we have obtained foundational understanding in light of all the above indications, verifications and evidence based on the amalgamation of science, reason, logic and facts as to who it is to be qualified for the title 'the creator of the heavens and the earth' and that being 'singular source' behind the universe.

Consequently, it only makes sense to agree that we ought to scrutinize the major religious systems' justification for their claim. The 'burden of proof' lies with the claimant who is asserting his/her 'doctrine' to the general population. Hence, we will examine the books which contain these religious belief systems 'theology', which is how each religious belief forwards its justification for its claim.

The caveat, however, is that this cannot be an exhaustive examination of each set of beliefs due to the enormous scope of manifestations required, however, and more importantly, due to our 'relative reality, ' that is, the efficacy required to form the facts with respect to these religious systems concerning our goal of verifying whether they conform to the veracity of 'Singularity' the single source by the beginning of our universe is sufficient to aim for and the best way to move forward.

With that said, the aim herewith is not to irritate the followers of these religious beliefs, nor to skew or defame their claims without any basis, the aim is to uncover the truth and perhaps address the fundamental question everyone is asking themselves 'subconsciously' which is taking into account 'the undeniable fact' that we as humans must come from the same source hence why are there various conflicting beliefs as to who that source is hence many incompatible religious beliefs are practiced each corner of the world claiming that they possess that source.

Besides, it is our Human Right (HR) to be able to scrutinize any doctrine which is for or found in the 'public domain' any philosophical premise that attempts to claim the purpose of our universe and existence. Consequently, examining that claim is the truest purpose of Freedom of Thought and Conscience as well as the Freedom of Expression and Speech enshrined in our Universal Human Rights Declaration (UDHR).

For those of you who are wondering what the former right entails, it is precisely to safeguard your right to change your belief system at will should you choose to do so without any external interference. Furthermore, it is the most substantively 'absolute' right of all rights due to its autonomous implementation because, after all, what belief you hold in your heart is literally 'yours' beyond any external coercion.

Subsequently, the immortal intelligence will play the shell game to uncover what is under the shell? Because every religious system is wearing the same coat 'shell' in the name of religion, consequently, by removing the shell, we would be able to ascertain whether this belief system is conforming to our objective of the Singularity or not.

The Immortal Intelligence
Plays the Shell Game

As established above, there is a singular source from which all that exists, whether solid 'matter', energy or moist 'water', stems as a point of origin. Consequently, the logic, reason and rationale which are the seeds of 'immortal intelligence' dictate that there is only one correct way to connect to that source, but the issue is there are plenty of philosophies, doctrines and religions claiming to possess that singular way hence a shell game puzzle to play.

The immortally intelligent people already comprehend this notion as a matter of 'self-evident' recognition . If there is a singular source and, consequently, one outlet for existence; hence, there is similar singular reciprocity. In other words, there is a singular entrance to that outlet, as referenced in our Big Bang section above. An example is conferred: if you want to enter the employment force of a successful company, there is only 'typically' one way to achieve that goal, which is to apply for it through the recruitment division of the human resource department of that company, and further to be admitted to join the workforce of such organization you must fulfill the requirement of such firm with regards to the post you applied.

As said above, a sincere seeker of the truth would face, along with defeating the 'artificial orbit' to make the time to activate his/her immortal intelligence, what appears to be the daunting challenge of sifting through all these organized religions and philosophies. Much like the shell game, the tactic to suspend your quest includes 'wearing you out' through such an effort, which often leads to the notion that if one organized religion or philosophy is bogus, then all are equivalent, all are bogus; dogmatic theology is the basis which they justify your allegiance to ignore the flaws, to reach to that end they introduce the so-called term of 'faith'. Which is the reason why the above description is given to it 'faith' because there is no substantial evidence, reason or logic which justifies our conformity to what is proposed to us.

If the afore scenario does not work with you, the usual tactics applied to you and me include it is a 'tradition' which diverts grounds of acceptance from evidence and reasoning to emotional blackmail: 'it is our culture' it is our identity, it is what your folks practiced, your ancestors held etcetera with repercussions as a consequence to object, question and demand intellectual justification to its core theology and practice. Such repercussions include labeling the immortal intelligent people with names like you would qualify to be heretic or blasphemer, you are an odd person out of the society, no less than an alien to your people are some of the things poor immortal intelligent people are threatened with in order to bully them away from activating their immortal intelligence.

But these are no more than scaremongering tactics we are accustomed to hearing every day to prevent us from succeeding. How many times have you heard your dream is impossible, out of your reach to accomplish, whether your dream is becoming a champion in sports, a scientist, the mayor of your local municipality or, the president of the United States or whatever else it might be including the next Steve Jobs of the Tech world. If Barack Obama were to obey such 'narratives', let alone becoming the president, he would not even have 'the audacity to hope' for his dream of running for the highest office in the land. The value of immortality by its virtue of 'time' alone, as the name 'immortal intelligence' suggests, outweighs all the above benchmarks; hence, this should be treated to be superior to anything else.

Another tactic to dilute your resolve to get to the bottom of these philosophies in your quest is all religions teach to do good, be nice to your neighbor and fellow citizen, hold the door for the next person, be good to the environment, assist the disabled and elderly people etcetera.

This narrative you might hear from many of your friends and colleagues. However, the problem is, these mentioned above are addressing the branches of the tree while ignoring the trunk of the tree. Only superficially addressing the symptoms while neglecting the source. Consequently, as such branches cannot exist without the trunk of the tree, thus where the trunk is unidentified, the branches become relatively insignificant or even irrelevant altogether. An example is conferred: Imagine you did all the general good conducts mentioned above but did not recognize your government's core obligation to pay tax; nothing of good conduct and behaviour would vindicate you from being liable to tax evasion, similarly just doing good, while it is attractive, is not addressing the fundamental pursuit to probing the 'big question' who invited us to be alive on the planet Earth? Therefore the immortal intelligence is transcendent to the branches and goes to the trunk, the foundation, the root cause of the issue.

Consequently, much like the 'shell game', your quest will be your participatory event of playing the shell game puzzle with these religions, which entails that you are aware of, on the outward appearance, these belief systems are all dressed alike in the name of religion, offer numerous etiquette traits. However, only one is containing the prize inside the shell due to the nature of the 'singularity' we are after.

Unlike the shell game, however, which is designed for us to lose because we have only one try at a time before we lose our money, the immortal intelligence's quest is penal-free for opening all the shells in order to authenticate which one has the prize, subsequently, toss away the irrelevant belief systems that fail to equate themselves with this Singular source of the universe as discussed above. You might want to remain 'stick around' for this until the end of the shell game, you might be surprised!

The probable ground why we have so many religions is to hide the truth under the heavy volume of information so that the seeker is inundated with vast materials to analyze and consequently 'wears off' over time and gives up his/her quest, which is something so many of us are caught with, this is tradition something 'open-secretly' practiced by law firms when disclosure of the information is requested from them. In this way, the party did not directly misinform but made the truth so difficult to be found by its seeker.

The Controversy of One Singular God
Why Many Religions

Many of us, including myself, always wondered if we, as humans, who are the same creation with the same requirements for life, such as oxygen, water, food, and the like, which means to the 'impartial judgment and reason' that we are all emanating from the same source, then what caused the proliferation of deities and gods? More relevantly, how did we get here with the current world's cluster of religious systems?

Further, how religions synonymously blended with our beautiful variety of cultures and food propriety, etcetera. Where religions are 'supposedly' to explain the source of our creation, ought not to be given an identical regard with cultural propriety, notwithstanding the two overlap and share common customary diligence; nevertheless, culture is reprieved from answering the big question where religion is not.

In other words, cultures do not have to explain 'how we got here', whereas religions must in the immortal intelligence approach; therefore, the two ought not be regarded as precisely identical. To further double-down the subject matter, if you can bypass the organized indoctrination of these various organized religions, and by that, it means you directly aim at analyzing their claimed books rather than spoon-fed by interpreters in clergy persons 'the middle guy', you would be astounded on the question of who is the source behind the creation conforms to the 'singular unseen' source we discussed above.

Furthermore, it is an even more astonishing 'fact' that an impartial judgment would arrive at a central more transcendent figure being the source of power and existence in all main religions of the world without exception. Whether we desire it or not, whether we are looking forward to it or not, the name 'Allah' often appears in different forms, adjectives, accents, dialects, etcetera.

Allah means the 'Singular God' in one word, the singular unseen source behind the creation, which is something we extensively covered previously in our discussion. This is not empty rhetoric, nor your everyday 'mythic narrative', not your conservative versus liberal discourse, far-right versus left and/or WOKE. It is not even in any ambiguous terms. Please read further to be astounded by the basis of this news acclaimed herewith.

Since major global religious belief systems root back to these two strands of Dharmic and Abrahamic religions, in some of the major Dharmic theologies such as Hinduism and Buddhism, the name Allah, 'the singular source', usually is embodied as a meaning which subscribes to the beginning was one singular source a 'Supreme Power'. However, conversely, the Abrahamic religions such as Judaism, Christianity and Islam, contrary to common misconceptions, the name Allah is a precise pronouncement with slight variations due to different tongues of people with their variety of languages and dialects.

As far as Dharmic beliefs are concerned, the Hinduism system, being the most major philosophy with the largest following in this strand, has retained books which are self-explanatory to the immortal intelligence, such authoritative religious 'scriptures' which declare the nature of this singular source of supreme power are as follows: In Chandogya Upanishad in Chapter 6, section 2, verse 1, this singular source is described to be "Ekam Evadvitiam", which translates to mean He 'the supreme power' is only one without a second, classifying the creator to be singular in its quantity, which unequivocally amounts to be antonymous to the multiplicity of gods in Hinduism doctrine practiced today. The book contradicts the organized, indoctrinated interpretation of Hinduism. Shvetashvatara Upanishad chapter 6 verse 9 which further verifies this being has neither parents nor children: "of him there are neither parents nor Lord".

Shvetashvatara Upanishad chapter 4 verse 20, this being cannot be visually seen: "his form cannot be seen, no one sees him with the eye". In Bhagavad Gita chapter 10, verse 3, "You should learn me as unborn, beginning less, as the supreme lord of the worlds". Further, the book Yajurveda (32:3), there is no image of this being, something we are familiar with in science, the Bible and the Koran as well. The list of evidence is far longer; however, our approach in this book has never been nor supposed to be exhaustive, but to present facts in a clear and concise way in order to assist the reader to see the truth beyond any reasonable doubt.

As far as Buddhism is concerned, as many people come to the same conclusion, it is more of a code of conduct than an engager with the big question as to who is the candidate behind the universe or how this universe all started. As mentioned above, the founder of Buddhism, Mr Siddharth, later known as Gautama Buddha, who is a learned scholar in both Hinduism and Jainism, at the age of 35 proclaimed the title Buddha, which means 'the enlightened one'. He avoided 'himself' engaging with answering the big question, even when directly hard-pressed to give some kind of response with regards to whether the Creator of the universe is the Singularity or Multiplicity. He deflected from engaging with it by giving a response as, "If you are suffering from a stomach ache, would you concentrate on relieving the pain or studying the prescription of the physician?" "It is not my business or yours to find out whether there is God? Our business is to remove the sufferings of the world".

Whether we admit it or not, the man Buddha withdrew himself from explaining our targeted question of who is the source behind the universe. Therefore, who are we to speak on his behalf to answer this question when he clearly says it is not my business to find answers to this very question? His teachings consequently amount to be a good conduct of ethical model seemingly equivalent to ethical models out there such as Nolan Principles comprising Selflessness, Integrity, Objectivity, Accountability, Openness, Honesty, and Leadership.

As such, therefore, the immortal intelligence concludes that Buddhism is no more than a set of good behavioral and moral conducts in order to attain humility and foster a good character out of its followers. It teaches models such as the Noble Eightfold Path, overarching Right View, Right Thoughts, Right Speech, Right Actions, Right Livelihood, Right Efforts, Right Mindfulness, and Right Meditation.

However, it lacks the title of being a full-fledged religion by our standards because Buddha himself refrained from addressing the big question. Hence, as such, a philosophy such as Buddhism, whose founder Buddha 'himself' refused to fill in the question, cannot be considered a candidate to the essential quest of this book. We will briefly engage with it further on the basis of our 'relative reality' relevant to us.

However, when it comes to the Abrahamic scriptures, not only attributes of this singular being are conferred, but his direct name is classified and declared. Nothing can come close to the name Allah in its different pronunciations of different Semitic languages such as Syriac or Canaanite, the language spoken by the patriarch Abraham, the name is Alaha; in Hebrew, the language spoken by Moses and many other prophets, the name is El, Elohim, Eloah, in Aramaic, the language attributed to Jesus to have spoken, yet the same name appears as Elah or Allaha. In Arabic, the most widely spoken Semitic language to date and the scriptural language which the Koran was revealed upon, it is also the language spoken by Ishmael, the son of Abraham and his descendants, including Muhammad of Islam, the very same name Allah appears yet again to be the singular source behind the creation of our universe.

As said above, the name **Allah** is the most predominant descriptive name given to the central power behind the creation, and it overwhelms any other names provided on the aggregate of **Abrahamic** religions. The **El** and **Elohim** alone appear over 2600 times in the Bible. Further, it is the personal individualistic name allocated for this supreme being in both the Bible and the Koran, '**Allah**', with the distinctive creative power behind the erection of the universe from a singular source, 'Singularity'.

In the Bible it teaches mankind in Genesis 1:1 that it is 'Elohim' who was alone to create the heavens and the earth; here, it is distinctive to clarify the singularity of its erection; no one else partook in this assignment except Elohim/Allah, neither Jesus nor Muhammad is mentioned here, not even other names of God is deployed for this all-important communiqué and function in the Bible. So, there should not be any needless confusion as this is pertinent to our central question of who is behind the Big Bang.

You do not hear YHWH, Adonai and so on in this verse. Despite this fact, the organized clergy in Christianity as well as Judaism, perhaps to a lesser extent, do not advocate this unique name of God Ela/Elohim/Allah. You 'almost' would never hear it in your weekly services on Saturday or Sunday for these two respective belief systems of Judaism and Christianity, except what is a slip of the tongue, etcetera. It would not be distant from the truth if 'the artificial orbit' intentionally overlooks its existence in the hope that the name 'Allah' withers away in old scriptural books without the public being informed about it.

However, the fact remains that it is less propagated than other names of God; as a factual result, if you wonder why that is, that makes 'the two of us' my immortal intelligence to wonder. The only plausible answer is 'unfortunately' it is all too familiar prospect to 'divide and rule' the general population of the world. To create the dichotomy such as 'them versus us' while astonishingly the underlying objective target has always been the same in the Abrahamic books, which is the constant, unchanging factor of Elohim/Eloah/Allah.

This is further corroborated by what would become known as the first commandment in the Bible, in other words, the most important requirement for the human who intends to benefit from his majesty's creation, such as water, oxygen and other sustenance provided, including fruits, vegetables and meat. In Exodus 20:3-5 which clearly commands its followers the worship of singular God in verse 3, "You shall have no other gods before me", which equates with precisely the first part of the Islamic covenant known as the 'Shahada' which directly amounts to be 'the witness statement' one gives in order to declare his/her allegiance to the same 'Singular Source' Allah "There are no gods except the singular God".

In verse 4 of Exodus continues that there ought not be any images of this singular power in the heavens and the earth. This rules out such pictures of the persona of Jesus or his mother Mary to be regarded as divine pictures, it is also extended to any additional persons blessed or not from prophets such as Moses and Muhammad to 'Saints'. The scope of prohibition also applies to angelic beings such as the Archangel Gabriel and Micheal: "You shall not make for yourself an image in the form of anything in heaven above or on the earth beneath or in the waters below".

The prohibition words of 'anything in the heavens and the earth' is the component that rules out the above persons and beings to be worshipped. Furthermore, you are banned from worshipping any being other than Elohim/Allah in verse 5: "You shall not bow down to them or worship them....." this prohibition of warship is blanket and applies to all, including Jesus and his mother, other blessed people such as Moses and Muhammad, it also applies to angels such as Gabriel and Micheal without exceptions and/or exclusions.

Let us learn together briefly how the name Allah is deployed in these Abrahamic scriptural books such as the Bible and the Koran that the organized clergymen would rather you and me not to discover, not to 'connect the dots' in order to see that we always had the same lord, the same goal, the same source behind the creation which is our current aim at hand.

As derived from the above scriptures without exceptions, the name Allah/El/Elohim is intended to serve a special purpose, a unique and singular purpose, which is simple in its understanding, so that people such as you and me who are plain 'truth seekers' not to be bewildered 'mixed-up' with this all-time great entity 'the Creator' with other created beings. Thus, this is the most 'effective' separation between 'creation and their Creator'.

The Bible Genesis 1:1 says it is rather Elohim alone who created the heavens and the earth at the 'Singularity stage'; the Koran further confirms in the clearest terms possible Koran 7:54: "No doubt your lord is Allah who created the heavens and the earth in six periods". This formula should be applied to any proposals out there as to who has the right of claim when it comes to this 'Singular God'.

An example is given: many people hold the person 'Jesus' to have the right of claim, as science has proven beforehand that at the emergence of the first gasses that formed the Big Bang from the singular outlet 'Singularity', humans did not exist at that stage, let alone Jesus and his mother Mary 'whose' womb he was born out of did not exist, could not exist, because the things that human life is dependent on to exist such as solar star like the sun did not exist at that time, so as planet earth and its life-supporting mechanisms such as vegetations, livestock etcetera.

This is further compounded by the Bible itself saying that Jesus is a created being, by who, of course, by God, and who is God, of course 'Eli/Eloe/El/Elohim/Allah'. The Bible Colossians 1:15-16 says Jesus is amongst the created beings, irrespective of whether he was the first or the last of the creation. Nevertheless, the bottom line does not change the fact that it says he 'Jesus' is amongst the created beings; this is not to reignite internal ancient debate within Christendom theologians and philosophers as to who is Jesus, is he Fully God 'Trinitarian' or fully human 'Unitarian' or even semi-god and human simultaneously, at which point is he God or at which point is he created human being, which books of the Bible should be kept in and given authority and which other books should be thrown out at 'will' in the council of Nicaea 325 years after Jesus, together with the council of Chalcedon, depending on which narrative the 'King' supports, but this is strictly to point out to the truth-seeking impartial immortal intelligence the ultimate reality, neither this is to cover up other passages within the conflicting Bible which upgrade Jesus to be Almighty God within the Bible, either way, you have a book 'Bible' which contradicts itself on the subject matter.

However, according to science along with the Bible 'put together' Jesus fails to have any claim to the throne of who is the Creator of the heavens and the earth, the father Allah/Elohim, however, has so far the most legitimate claim to this unique position. If there is still any doubt in your mind, consider it from the mouth of Jesus himself.

According to the Bible, Jesus is humbly crying out 'loud voice' to the Singular-Creator, the father Allah/Elohim, the God and says in Mathew 27:46, "My God, my God, why have you forsaken me?"therefore, Jesus has a God without any doubt what(so)ever as he says 'my God'.

Further, Jesus did not humbly cry out in the English language, according to this passage of the Bible. Therefore, what did he call God the Singular Creator from his own mouth, 'Eli, Eli, ' which complies with the different ways of saying El/Allah/Elohim/Eloe as it is exclaimed from the very mouth of Jesus to be his God? The same meaning is repeated in Mark 15:34, in which Jesus calls his God "Elo".

Hence, 'Eloi' which takes, therefore, the meaning of 'my God' as the passage informs us. If, after this, you allow the 'artificial orbit' to bamboozle your intelligence to cover up your eyes, you somehow are still surrendering your reason and independent thinking capability to 'blind faith' that Jesus is the Creator of the heavens and the earth. Then, my dear reader, realize that it has always been 'on you' up to you of your free will to decide what is best for your soul in your future.

However, let us shed light on the reasons which make 'the artificial orbit' to sell both concepts of 'Under-Grace' and 'Random-Chance' rather with ease upon large portions of the global population because both of the above concepts have the same common attraction 'feature' which is 'impunity' 'free lunch' which means the absolute absence of accountability.

It is all too common reality that almost everyone who is scammed or deceived by a third party diluted his intelligence either partially or totally prior to the trickery and/or the loss taking effect on them. Their intellect turns a blind eye to the trap set before them because their impulsive desire sees the utopic target they were allured into attaining.

An example is provided: Many people get an advertisement through the mail which tells them that they have already been selected, that they have won a brand new car or Lottery, etcetera, but they have to pay a processing fee in order to finalize their win. The scammer appeals to the human's weakness of strong longing for wealth. Not only that, the enjoyment and freedom that comes with the 'winning' money is the specific target for this allurement, the frequent vacations which become readily available due to the 'win' becoming wealthy person amongst endless additional privileges that are tied to winning are also the picture the poor scammed person would depict. It is no surprise that the scam works 'almost' always by virtue of its perpetuating existence and practice on people.

The notion of 'Under Grace' enshrined in the doctrine of Trinity comes on the backdrop of a 'free ticket' to a utopic prize of Gardens of Eden, except there is the requirement that you have to believe in blind 'dogmatic proposal' that God had condemned his only son to punishment in order to pay your own wrongdoings so that you automatically qualify to this all-time enjoyment of paradise which means the believer of this doctrine now lives and/or exists under the concept referred as 'under grace' it means in our lingua franca, life with 'impunity' or the absolute dissolution with regards to accountability which renders religious duties to suddenly become inconsequential.

Albeit, the common person is allured with the utopic longing naturally attached to human hearts; as such, the deceived laxes the full deployment of his/her probing capability to ascertain this all too-good-to-be-true offer, we called it the 'defensive division of intelligence'. In this case, the average believer of this doctrine never asks for confirmation, which is the least act required from any reasonable person. The deceived takes this doctrine as a capsule without any information label attached to it as to what its content is inside it. Pursuant to the subject matter, the deceived receives this free ticket 'under grace' with an inconsequential 'non-effort requirement' lifestyle as its unique selling point (USP) to attract its followers, which in turn kills the effort-based lifestyle of people that creates superiority predicated on action and/or effort from individuals.

This concept of 'under Grace' mimics the ideology of Marxism which also suspends the efforts of hard workers who shall be rewarded 'hierarchy' according to their efforts. Under this doctrine, whether you are a compassionate, charitable human being or a serial killer 'bloodthirsty' evil person, you would be repaid equally under God as long as you believe in this doctrine that someone else suffers for your sins, which also serves there is no point to the creation.

As we said above, the 'artificial orbit' sells this concept with ease to common people, while people all they see is the utopic end painted to them by prominent Bible writers such as Saint Paul, who reprimands anyone who places this doctrine under scrutiny as he calls them 'fools' for instance in Galatians 3:1 'you fools 'people' of Galatians, who has bewitched your eyes, don't you realize that Jesus Christ was killed 'crucified' under the 'cursed law' where its constant observation is repealed 'nullified' by his death on the cross 'paraphrased'.

Within the Bible, Jesus Christ, whom the religion is named after 'Christianity', unlike the man who called himself Saint Paul does not agree with the above doctrine taught by Saint Paul and his aiders. According to Jesus, there is no room for 'under grace' unaccountability, impunity or a free ticket to God's kingdom of heaven as we shall be rewarded according to our strife and effort. Jesus says with great emphasis in the Bible in Matthew 5:17-19 in verse 17 do never think I came to abolish the law of Moses and prophets 'the commandments' which directly contradicts and goes against Saint Paul's doctrine of 'under grace' which by all intents and purposes teaches the abolishment of the commandments. Jesus continues in verse 18 to tell us that this law that 'the artificial orbit' wants to abolish through 'Under Grace' will never be dissolved with perpetuity time clause 'until the heavens and earth disappear', not a dot or title, not a comma or period will be removed until its goal of practicing is achieved.

According to Jesus in continuation of Matthew chapter 5 verse 19, anybody upholding or teaching 'under grace' would not be acknowledged in the sight of God as Jesus says,"Whoever therefore breaks one of the least of these commandments, and teaches men so, shall be called least in the kingdom of heaven; but whoever does and teaches them, he shall be called great in the kingdom of heaven" As you can see herewith Jesus links one's effort to uphold the commandments of the Old Testament is not only the requirement to attain Eloi/Elohim/Allah's acceptance but there cannot be any other way to do so.

Thus, 'under grace' lacks the merit in achieving the kingdom of heaven as it were, which means, in our contemporary vernacular, a 'paradise'. Notice, by Jesus emphatically expressing whoever 'teaches man' to break these commandments 'under grace' is in breach of 'Christ' himself as he calls that person an offender against the Elohim/Elah/Allah when he 'Jesus' says would amount to be 'the least in the Kingdom of Heaven'. Therefore, it is crystal clear that the man called Saint Paul would qualify to be in breach of this verse due to the fact that he teaches the abolishment of the commandments and, as such, therefore, according to this verse, would amount to be the least in the kingdom of heaven.

If, after the above evidence, you somehow want to still follow the all 'too good' baiting allurement of 'under Grace' by Saint Paul against Jesus himself, then, my dear reader, it has always been 'on you ' with your free will to choose what is best for your soul.

What one can do with guilt-free feeling or without contempt in the 'under grace' doctrine cannot be achieved through Jesus' upholding the commandments. An example is given: in the under-grace doctrine, one can go to a restaurant and eat BBQ pork chops with his friends and family, whereas under the Jesus-affirmed law of the prophets cannot. In Leviticus 11:7-8 pork is unclean and not to be eaten. Similarly, the same prohibition is repeated in Deuteronomy chapter 14, verse 8.

Therefore, if you happen to be a frequent eater of bacon and call yourself a Jesus follower, please realize you are in major breach of Jesus' Christianity, which calls for the differentiation of Jesus' Christianity from Saint Paul's doctrine. Neither going down to the pub or bar with friends and family and start drinking intoxicants is recommended according to the laws affirmed by Jesus in the commandments, the Book of Proverbs 20:1: "Wine is a mocker, strong drink is raging: and whosoever is deceived thereby is not wise". Parallel caution is rendered in Ephesians chapter 5:18, which commands us to not get drunk as it leads to unfavorable results and debauchery.

Now you can see why those who frequently are longing for the above two indulgencies are inclined to follow Saint Paul's doctrine against Jesus' guidance and prescription but ironically in the name of Jesus because the whole belief system is claimed in the name Christ in Christianity, not Saint Paul.

Without any exaggeration whatsoever this notion of warship and practice that was advocated by Jesus has died in today's Christianity as Saint Paul won against the teachings of Jesus; however, astonishingly, Jesus was right with his perpetuity time scope of 'until the heavens and earth disappear' the law would be upheld. Although Saint Paul was fairly successful in eradicating this call of Jesus within most Christian churches of today, astoundingly, the law of prophets with their 'commandments' are fulfilled until the heavens and earth perish, meaning 'forever' through the Koran, the Last Testament, the youngest and most updated 'uncorrupted' as it calls 'itself' of Abrahamic beliefs.

If you cannot believe the above statement, yes, the ultimate truth of God 'Eloe' has its own 'mysterious way' of surprising people, and therefore, you and I are not the exceptions either. We will briefly engage with the battle between Jesus' gospel versus his opponents' within the Bible further under the heading of the Immortal Intelligence Engages with the Bible. However, the obvious disclaimer is that this book is not specifically about the authentication of the message of Jesus within the Bible.

The notion of 'Random Chance', too, removes the sense of purposefulness and responsibility from the creation. If the planet Earth popped-up without any reason or cause behind the effect, then nothing can give rise to any religious obligations, including the essential recognition for the Singular Source of Power who made it possible. An example is provided: imagine you found an envelope which contains $1000 cash inside it at your workplace. If you make the correct deliberation that this money was likely dropped by someone, perhaps a fellow employee in the company you work for;

consequently, this premise of thinking will trigger you to be responsible; thus, you would hand over this envelope to the lost and found department of the company.

However, if you are longing to take it with a guilt-free feeling, you would easily believe that this money popped-up in the workplace all by itself, 'a random chance' of a sort. Henceforth, this will not trigger 'any' sense of responsibility or purpose. In fact, conversely, it may foster its founder to establish the premise with the intent to take and utilize the money without feeling any remorse about the person who may have dropped it. For this reason, those whose concept of 'freedom' entails an unpragmatic fairytale of absolute deregulation would opt for this premise of 'Random Chance'; the universe popped-up along with everything in it, including you and I, without any cause or reason, enjoy it as much as you can, there is no any other point. This creates the debate between the romanticized fairytale of freedom versus the real freedom based on the circumstances on the ground.

The Ultimate Meaning of Freedom

Reverting to our question, given the above clear evidence beyond any reasonable doubt that all the founding persons in the major religions of the world, including Moses, Jesus and Muhammad, as well as the text of Hinduism as referenced above, pointing out a higher singular unseen power with the name of Allah in different variations, then how come there are too many religions which amount to colossal disproportionate to 'the singularity' and how freedom is linked with this.

In so far as the scope of this book, the simple answer is two factors played crucial roles in creating this proliferation and global divisions when it comes to religions. Not only that, further there are deviations and sects within each religion, constantly multiplying even further as time goes by. These two factors are 'power and wealth', religion represents to many both these attractive instruments, and the seekers of both have utilized religion to reach that end.

Contrary to common misinformation, the religion preached by these above founders created the notion of pragmatic freedom in which people/followers rely on the creator and consequently maintain essential independence from the creation. This proposal represented a threat to the hegemonic individuals, organized groups or 'kings' and tribal systems which had a firm grip on their citizens or, as it were, their subjects, which justifies the reason why these founders went through early struggles of 'talking truth to power' without exception, this frequently placed them under persecution and torture not only 'themselves', but their early followers as well.

Moses faced the tyrant Pharaonic King Ramses II who used religion as a vehicle to retain power and increase wealth as he conjoined religion and secular control as one paradigm over people by claiming to be the God Almighty himself. This claim of Pharaoh is different from most other power-hungry tyrant kings of the past, where other kings maintained to be separate from God Almighty while using the religious temple and its clergy to be their proxies and thereby interpreting religion for the interest of the King, the Ramses II conjoined the two where the king has become the 'God' Almighty himself and by extension the religion and the religion has become the king.

This empowered him not needing the clergyman to interpret religious text and, where necessary, create a religious doctrine which advances the policies and interests of the king or, as we call it today, the 'State'. In this way, the king eliminated the 'middle-man'; while it sounds like a smart move on his part, it is nevertheless a flawed political strategy because people are always as smart as each other; that is, Ramses II would not possess supernatural intellect over the rest.

By that, people are connected to God through the superior relationship based on nurturing and love because the One-true-God does not take from the people but rather gives, nurtures and provides for them. On the other hand, any human king 'tyrant or not, ' is taking from the people, seeking labor, tax and wealth from his 'subjects' which most of the time resulted in an iron-fist power control sustained through fear and cruel punishment of forced labor and slavery on the part of Ramses II in particular as he subjected people of Egypt mainly early Israelites under such treatment similar to 'in some respects' the slavery practiced in the USA slave trade between 16th to 18th century.

As Moses said to his followers remember the freedom granted to you by your lord 'Elohim' as he delivered you from the 'house of bondage' the land of cruel slavery in Exodus 13:3. Here, freedom is presented as 'unconditional' remembrance and servitude offered to the Elohim/Elaha/Allah in lieu of tyrants like Ramses II. In other words, serving Allah alone and directly without any preconditions is the pragmatic freedom itself in plain language, a paradigm we will elaborate further in later stages.

On the other hand, Jesus who sought freedom for his people, who were under the control of organized religious clergymen known as Pharisees and Sadducees, a 'political wing' for generating subscription payments from the unlearned public through the medium of placing themselves between God El/Elohim/Allah and the common people in addition to the pagan polytheistic Roman emperor Tiberius Julius Caesar who controlled the Holy Land of Jerusalem and its vicinity, the entire Levant area were under occupation at the time.

Although Jesus was not successful in reaching this freedom like Moses evident in his court appearance on the so-called Good Friday in which he denied the two counts of charge, the first being are you the Son of God and the second being, do you claim to be the King of the Jews; to both counts, Jesus responded "you have said so" i. e. 'not me' when the Governor Pontius Pilate who presided over this court realized that there are no merits to these charges following the earlier rejection of previous judges such as Sanhedrin and Herod Antipas to hear this case, he symbolically requested the service of the jug of water and washed his hands before the eyesight of the chanting crowd of Jesus' enemies 'Pharisees' to demonstrate that he had nothing to do with this 'bloody' injustice. Henceforth handed over the judgement of Jesus' fate to the very same Pharisees, who invented these baseless accusations in the first place due to their coercive lobbying tactics. As a reference, in Matthew 26:57-27:31, Mark 14:53-15:20, Luke 22:54-23:26, and John 18:13-19:16.

Subsequently, the foregone verdict of 'crucifixion' was readily passed with ease to be commenced upon him 'Jesus', which by its success would bring demise to his presence and message on the whole due to the fact that the people whom he is seeking freedom for, from the corrupted temple would not accept him as 'the Messiah' being a word of God who can be crucified, the reason being is that the very commandments preached by Jesus clearly state that he who carries a capital offence punishable by death and therefore crucified 'hanged on a tree' is cursed by God in Deuteronomy 21:22-23.

His people understood you cannot be the word of God, the 'Messiah' on the one hand and be cursed by God on the other hand like a common criminal and crucified accordingly, which is the catalyst factor that 'Judaism' holds this view 'to date' that the person who had feuded against the Pharisees was not Jesus even though it was him indeed Jesus the Messiah because had he not been 'the real' Messiah all the supporting evidence with respect to the number of miracles performed by him must also simply amount to lies and baseless which cannot be.

What these opposing premises show is that Jesus the Messiah as 'a word of God' could not have been crucified according to the above evidence in Deuteronomy 21:22-23, which is the position held by Judaism to date. Further, the person who committed a priori 'well established' high number of miracles, which enjoys uncountable 'mass reporting' across all Abrahamic scriptures, including the Koran, is not Jesus the Messiah, is also a position held by Judaism which is illogic, weak and nonsensical.

While in multiple places in the New Testament, such as Matthew 27:32-33 and Mark 15:21-22 inter alia, many more parts of the New Testament state that Jesus the Messiah, the word of God placed on Mary was crucified is a position held by Christians in contradiction with the Old Testament as referenced above which is the point of contest between Christianity versus Judaism.

The immortal intelligence sheds light on the interesting Koranic account of this long-standing disagreement between the New and Old Testament of the Bible, i. e. Christians versus Jews. Which is it was the real Jesus who committed all the 'mass-reported miracles' with the extensive supporting pieces of evidence. However, Jesus makes a special prayer to Allah to save himself from this embarrassment of crucifixion in Matthew 26:39, in which he prostrates to Allah and said if you would accept, can you save me this cup 'crucifixion' based on your will, not mine "He went a little farther and fell on His face, and prayed, saying, "O My Father, if it is possible, let this cup pass from Me; nevertheless, not as I will, but as You will. "

And as a result of this prayer, Jesus was spared from such an embarrassment and miraculously lifted up into the heavens to Allah in Koran 4:157-8. Jesus was not killed or crucified. Rather, it made appear to them that way, Jesus was raised to Allah. "And for boasting, "We killed the Messiah, Jesus, son of Mary, the messenger of Allah. "But they neither killed nor crucified him—it was only made to appear so. Even those who argue for this ⌐crucifixion⌐ are in doubt. They have no knowledge whatsoever—only making assumptions. They certainly did not kill him. Rather, Allah raised him up to Himself. And Allah is Almighty, All-Wise. "

As said above, Jesus wanted freedom for his people, which is untempered direct servitude and relationship with the creator; in our contemporary vernacular, no middle man, no agent, and no transit stops anywhere before serving God 'the Father' Elohim/Alaha/Allah, which was detrimental against Pharisees and Sadducees' financial income and power generated from the control of the layman masses.

As Jesus frequently used to emphasise this point to layman masses, in order to attain a successful afterlife 'the eternal life' you are required to give sole servitude to the Creator Elohim/Alaha/Allah and recognize Jesus as a sent representative of the Creator in John 17:13 "And this is eternal life, that they may know You, the only true God, and Jesus Christ whom You have sent", notice from the mouth of Jesus in this passage he 'Jesus' is separate from the Creator, he is sent by the creator, which amounts to consequently he 'Jesus' **is** not the creator.

Notice also that this statement and others like it, which were preached by Jesus, would effectively dismantle the middle power dependency created by Jesus' foes, such as Pharisees and Sadducees, in order to have a firm grip on the layman masses who were mostly illiterate in this era of time.

This, in turn, created the motive that Jesus must be stopped in his 'infancy' at an early stage of his mission before his teachings would take substantial traction among **the** masses, consequently the inception of a hostile united front from the established clergymen against Jesus in Jerusalem and its surrounding villages at the time. One of the tactical strategies employed was to implicate him in crimes that would carry the death penalty in that jurisdiction, which would serve the purpose of ending **his** mission in order for the established temples to continue to enjoy the power over the unlearned layman masses of Jerusalem and the surrounding villages.

Such implicated death penalty crimes were developed to be the two charges mentioned above, which contravened both religious and secular powers who were in control at the time. These 'high priests' who were the enemies of Jesus, were strategic with their accusations which they waged against him; the charge that Jesus is claiming the 'literal Son of God' would dismantle the spiritual legitimacy of Jesus within the Israelite community as this amounted to blasphemy in the Torah, while the second accusation charge; 'King of the Jews' would attract the wrath of the Roman power who were the occupying power of Jerusalem and its surrounding villages at the time, if this is unchecked he may mobilize and cause a general uprising against the occupying power.

The aim of this book is neither the history of Jesus nor was it designed to be as such. There are many works conducted with that aim in mind out there for those who would like to learn the historical events of Jesus. However, we have covered only the **areas** relevant to our quest of finding a sensible answer to the big question of who is the likely candidate for the position of the singular source behind the creation. The freedom Jesus wanted for his people was to declare independence from intermediaries, the 'middle guy', via giving unconditional love and servitude to the creator Elohim/Allah/Allaha, just like Moses before him.

Muhammad of Islam was no different. He wanted the same freedoms as Moses and Jesus for his people who were placed under 360 Gods along with their conflicting demands from the people. The 'Quraish' tribe who was in charge of the valley city of Mecca were the intermediaries 'middle guys'. Mecca was known as the house of the singular God of the Abrahamic religions. In the Bible, Mecca is referred **to** as the valley of 'Baca' in Psalms 84:1-12.

Astoundingly, this verse recited by the Messenger of God, King David, says, "......blessed are those whose hearts are set on pilgrimage as they pass through the valley of Baca.....**"Are** you planning to perform the Hajj pilgrimage to the valley of Mecca 'the largest and the longest ritual of the planet earth given to none other than the 'Singular God' to include yourself amongst the blessed people King David is referring to?

Reverting to our task at hand, given the ramifications of the new societal discourse debated and expressed by Muhammad in Mecca in the 6th century, which sought simply independence from the 'middle guys' with their countless 'idolatries' who represented the elite 'Quraish tribe', the financial and hegemonic influence over the people became an injurious threat and challenge to the monopoly of public narrative retained by the 'Quraish' under the polytheistic system.

It is quite interesting how Muhammad, the prophet of Islam, propagated for freedom of expression and opinion as he faced suppression and subjugation due to his teachings **which** were increasingly winning the hearts and minds of the public at the time. Furthermore, his early followers were persecuted, killed and tortured for the purpose of exerting deterrence upon the society so that it appears to the layman masses following Muhammad and the path of becoming his followers brings hardship. Despite this concocted effort, his followers kept on increasing and ironically **keep** on increasing to date even after his passing some millennium and a half ago as the fastest growing belief system 'religion' in the book Guinness as of 2024 not only due to high fertility rate but also high intrinsic appeal to many.

It is remarkable how Muhammad sought political asylum from neighboring cities 'States' as it were due to suppression of his freedom of expression and opinion in his own town of Mecca 'Makkah' something enshrined in the Geneva Conventions of today and theoretically International Human Rights Law such as the International Covenant on Civil and Political Rights (ICCPR) because it lacks the proper enforcement in some places.

Prophet Muhammad used to plea to the chieftains and elders who controlled neighboring cities and say, ". . . My people prevented me from freely expressing the truth of 'the singular God', will you grant me protection in your city?" This later trickled down and celebrated in the concepts of some of the fundamental freedoms many advanced nations cherish today, as many philosophers also argued as a precursor to modernity 18th century, such as JS Mills 'The marketplace of ideas', which means protection ought to be afforded to free and opposing ideas in the public sphere conversations as much as 'free enterprise' is granted protection, due to the free competition arisen from these competing ideas would result in the greatest civilizational achievements to humanity. This has become a fundamental block to proto-democracy, something barred today in most of the places where Muhammad had lived and preached to date, quite the antithesis of his ideals.

To Muhammad, freedom was not given to him on a silver platter. It neither was a gift nor a charity; he and the freedom-seeking people had to go through existential wars to achieve it, a real struggle; according to Muhammad, the proper use of the word Jihad 'a perseverance' the struggle to achieve good ends such as freedom is essential to human life. Some of these life-threatening wars he underwent included the war known as Mount 'Uhud' in which he was physically injured and as the enemy army was closing in on him and his face was bleeding due to his injuries, he exclaimed paraphrased 'Woe to, shame on these people trying to kill their prophet, how [on earth] will they be granted an entry to paradise" and later Koran the book he claims he did not author himself, rather it was revealed to him by Allah replies to him in an unfavorable reprimanding tone in chapter 3: 128 "Not for you, [O Muhammad, but for Allah], is the decision whether He should [cut them down] or forgive them or punish them, for indeed, they are wrongdoers" which says to him, it is not your privilege to decide who God condemns to hellfire and punishes or who God forgives, in so far as you are concerned yes they committed transgression and injustice against you which reinforces the claim that Muhammad is only a servant messenger not God himself.

The **Koran** reprimanding Muhammad and exposing his human weakness can be witnessed at will and cited in multiple places within the **Koran**, for instance, in Chapter 80:1-12 of the **Koran** as the prophet Muhammad was sharing his ideals of freedom with one of the elite personality in Mecca, a poor blind man approached him and asked him what is your message. Hence, he 'Muhammad' turned away from him in a frowning manner.

Hence, Allah in the Koran says in this chapter you 'mistakenly' disregarded the poor, impartial, sincere heart of the blind man in favor of the rich elite influential person, while the poor blind person is sincere and intent on seeking freedom and purification from these manmade fake gods, the rich and elite may reject 'your call' due to 'arrogance and pride' in his heart arising from his illusive temporary power and stature that would deplete at his death anyway. Koran 80:1-3 "He frowned and turned 'his attention' away, ˹simply˺ because the blind man came to him˹interrupting˺, You never know˹O Prophet˺, perhaps he may be purified. . . . . " These verses and the like astonishingly yet add to the validity that Muhammad could not have 'himself' authored this book called the Koran.

Muhammad, in the end, would triumph against his adversaries and draws a new historical chapter of forgiveness and the 'rule of law' governed upon people. In his seminal celebrated farewell speech in the 6th century, he outlined the notion of human equality he envisioned for humanity, that human beings are equal under God.

Not only that, he incorporated a brotherhood and sisterhood to human beings as he reminded us we all trace back to the same parents 'Adam and Eve' hence are 'Adaminians'. In that speech, he cures 'terminates' by way of abolishing the prevalent practices of the 'chain of being' class and racial superiority over other people, which promoted perpetual hierarchical authority of class of people over the others in the society.

Kings, Nobles and Lords have the moral ground to rule over the Commoners, Peasants and Plebs of the lower class of society, which we would loosely call the 'blue collar' class of today perhaps applicable to you and me. He particularly places an emphasis on what the immortal intelligence recognizes to be 'religious-racial superiority', which is when the followers of religion lay claim to ethno-racial ownership to the servitude of the 'Singular-God', the lord of the universe.

Such an example include 'Hinduism and Judaism' simply being born as a Jew or Indian race/ethnicity you are entitled to superiority over the Gentiles as he says no Arab should ever feel superior to others 'Ajabs' which translates to mean equivalence of Gentiles, in the same paragraph he aborts 'vaccinates' by way of prohibition against racial superiority no white person is superior to a non-white person, to Muhammad superiority is effort-based as such the hardworking person towards his Lord may attain superiority over the lazy working person, his speech reads "All mankind is from Adam and Eve, an Arab has no superiority over a non-Arab nor a non-Arab has any superiority over an Arab; also a White has no superiority over a Black nor a Black has any superiority over a White except by piety and good action. . . "

Further, Muhammad laid the foundation of human dignity of what would develop to be known as 'unalienable rights', the constant human rights that are not lost or gained due to who holds the power of the government at any given time. As Muhammad envisions within his speech to the well-being of humanity, three Human Rights are not to be infringed upon what he calls 'the blood, the property and the honor' of fellow human beings.

These rights would develop to become the right to life, owning of property and protection of baseless defamation including the right of private and family life. Unexpectedly, at the governmental and institutional levels, these rights and visions of Muhammad are better observed in the Western civilizations than in the East, something the immortal intelligence notices, which means if you live under the freedoms and protection of human rights in the West Muhammad's impact on your well-being is a reality that you may be oblivious to while you possibly maintain an antagonistic view against him without authentic knowledge of him. We should learn more about this man in the West! Furthermore, remarkably, Muhammad predicts, ". . .

the last ones may understand my words better than those who are listening to me directly".

Additionally, Muhammad's vision of civilization and equality of people would play a crucial role in stipulating and canonizing the firm erection of freedom of thought and choice. He would stipulate no one without any derogation should be forced into any theological doctrine, even if that ideology the person intends to reject is the truth, as he would recite the Koranic verse of 2:256, "Let there be no compulsive force in believing religion so that the truth is separated from falsehood. . . . ", simultaneously he would extend such freedoms to women's rights paraphrased "no young girl could be forced into a marriage that she does not consent. . ", to Muhammad, he lowers the age of consent to teenage girls hence to 'young people'.

These provisions of tolerance predicated on the right to possess different political views within the society as well as the courage to tell the truth to power would attract worldwide attention later during the Enlightenment era in particular, great philosophers such as John Locke and Jean-Jacques Rousseau, who played a crucial role in revolutionizing the aristocratic totalitarianism prevalent in Europe in that period utilized Muhammad's envisioned values and ideals.

John Locke, in his seminal works in his 'Epistola de tolerantia', a letter Concerning Toleration in October 1689 which was designed to speak the truth against power and further attributed to the advancement of modernity in the Western human rights and democracy writes paraphrased 'the business of religion is not to be instituted the exercise of compulsive force. . . ' in paragraph 7 echoing what Muhammad bravely said in the 6th century not only to his opposers but also to his followers.

In his famous 'Du Contrat Social' the social contract in 1762, Jean-Jacques Rousseau says a liberty for someone is 'obedience one gives to the laws one accepted for oneself'; he underscores the underpinning cause of 'choice' one has to make to follow what he/she already accepted. It is important however to make the distinction between majority and correctness.

Unlike Rousseau, Muhammad's ideals give high regard to the protection of freedom of choice on individual and/or collective levels regardless of general societal consent or not, as he 'himself' was under the persecution of the elite Meccan governing social committee who reached 'unanimous consensus' to eliminate 'murder' him due to his belief and ideals.

Unfortunately, this is not a book about the history of Muhammad but instead finding answers to the big question who is the likely candidate as to the creator of the heavens and the earth. However, there should be a more analytical review of Muhammad's contribution to human civilizations.

The pragmatic genuine freedom proposed by influential personalities such as Moses, Jesus and Muhammad was to relieve the burden of 'intermediaries' who conveniently placed themselves between the masses and their God, hence gained control, power and wealth over the general population, which is the stark contrast to the kind of delusional freedoms that the 'artificial orbit' misleads people with. To these above personalities, it is obvious tacit that if God 'Allah' is the source of life, to gain freedom from God is not a realistic or strategic goal, it is to 'cut off' the supply of life that God provides such as oxygen and water amongst additional necessities, which means to annihilate oneself.

An example is conferred: imagine a child is bamboozled to disconnect in the disguise of 'freedom' from his/her parents, who are the providers to that child. Should the child's wish be granted, then the child will no longer receive support from parents and hence would experience starvation and certain death. Not only does it make no sense the desire to be free from God, but it amounts to plain 'stupidity'.

It is no different than to fancy to be free from your bodily organs, such as your legs which provide you with the strength of movement. Alternatively, the wish to be free from God is equal to longing to be free from your own brain; rather, it is due to your brain that you are a dignified human being. Consequently, lack thereof would equate to other animals or cause fatality for your survival to exist as the medical definition of death is brain death, 'The irreversible cessation of all brain activity '.

The Immortal Intelligence Requires Primary Evidence 'Data'

As said above, now that we are pretty certain there is an unseen singular power behind the creation. Consequently, the immortal intelligence plays the shell game when it comes to the quest of this 'singular source of power'. Not everyone 'religion' who claims they have the winning prize under the shell actually can prove what they claim. Therefore, the immortal intelligence requires clear and coherent factual evidence that can be proven now to us in order for the religion to be accepted for its objective claim.

Just as researchers and institutions, including our legal courts, created a logical system to analyze and ascertain the viability of evidence henceforth adopted 'the admissibility standard' before any claim is accepted as evidence. Similarly, the immortal intelligence will deploy an effective standard fit for our investigation. Religions that only rely on secondary evidence and dogma with the complete absence of primary evidence would fail, whereas religions that can justify primary evidence and then, in addition, utilize secondary sources may become candidates in claiming 'the singular God'.

Henceforth, the immortal intelligence defines what constitutes primary evidence and what constitutes secondary evidence. The primary evidence in our investigation is the miracles that these books put forward that are either accessible to us today, or we can be eyewitnesses to them.

On the other hand, secondary evidence is miracles that these religions claim which are neither accessible to us today nor we can witness them directly. Further 'dogma or faith' is the reasonable belief of future events that we will have to trust that God will carry out provided that there is primary evidence to back it up the faith, just as your political party 'manifesto' which you are voting for the election year, you are trusting the leader of that party with his/her 'power' to be able to carry out what they promised.

An example is provided: religious miracles such as Moses splitting the Red Sea or, Jesus healing a blind person, or Muhammad splitting the moon may not constitute as primary evidence because we cannot access to be an eyewitness to these claimed events and consequently are inadmissible on their own. However, a clear disclaimer, that is not to say they did not occur, but the immortal intelligence classifies them as secondary evidence which can be used as supporting evidence provided there is primary evidence for them. Whereas scientific evidence referenced in any of these religious books that could not have been available based on the 'technology of the day' to any of these central figures and to these scriptures can qualify to be admissible as primary evidence because it is something we can access and/or witness today.

Likewise, any scientific or historical 'grave' mistake and/or errors beyond the semantics of the language would disqualify any of these books from being a candidate; just as any heavy contamination of food substance would alter the authenticity and purity of that food, grave mistakes and errors would also change the authenticity and the purity of the message these books are claiming to transmit.

The Immortal intelligence would only give consideration to limited religious belief systems due to our relative reality. These are Hinduism, Buddhism, Judaism, Christianity and Islam. If you would like to test other religions you can apply the immortal intelligence tools to verify if the religion is representative of the Singularity source in our universe without substantial contaminations. Furthermore, religions that lack this unseen power of Singularity would be disqualified.

The Immortal Intelligence
Engages with Hinduism

As hinted earlier, the immortal intelligence finds the name of the religion being 'Hinduism' disqualifying, in that, it cannot be representative of the transcendent singular creator of the universe. The word 'Hind' means and is the direct equivalent of the word 'Indian' in eastern cultures, which refers to a specific race/ethnicity of people who live in a specific particular area of land in Southeast Asia. To suggest that the requirement of this 'Singular' power is to be a 'Hindu'—an Indian, a specific race that people are born into defies the logic and reason which justifies our grounds for rejecting such a claim. You are born into a race and ethnicity; therefore, you do not choose to become a race. This is precisely why racism is deemed to be a 'crime' in most of the civilized world's legal systems.

Therefore to suggest that the singular power who created the universe is a racist by default is fantastically ludicrous. That said, the immortal intelligence must be fair to 'Hinduism' in recognizing that their earlier scriptural books did not call its theology as such, nor did the followers of this belief system. The name is suggested to have its roots in foreigners who visited that land, naming it after the source of water emanating from the river Indus within this territory.

Despite some 'Hindu' scholars providing limited rebuttals and discourses to counter this racial description, nonetheless it is fair to say that most followers of this faith system accept this nominal description given to them by foreigners for a good reason.

Not only do Hindus cannot agree on who God is, but they also lack consensus on how many gods there are to be worshipped. Consequently, nothing else can unite them other than the geographical and racial identity which they happened to have in common and share—'Hind/India.' Hence, the followers worship conflicting numbers of gods, ranging from three to a thousand and even 33 crores (330 million gods and deities), as referenced in the Brihadaranyaka Upanishad 3.9.1-2. The dispute over how to interpret this does not change the fact that there is a plurality of gods and deities. This plurality contradicts not only the well-established singularity of the power behind the universe but also the 'Singularity' enshrined in other Hindu scriptures as well, such as the concept of 'Ekam Evadvitiam' "He (God) is only one without any addition, one without second." in Chandogya Upanishad, Chapter 6 Section 2, Verse 1.

One methodology by which this plurality and multiplicity of deities and gods infiltrates human intellect is through the concept of 'Pantheism.' In effect, this creates nothing more than 'intermediaries' between the real singular creator and people. We should not entirely underestimate how this concept, with relative ease, bamboozles and 'hacks' the human brain's firewall of reason and logic. It trickles down even to the most monotheistic Abrahamic indoctrinations, including certain Muslim subsect communities.

People's intellect cannot be undermined by the objects they are told to offer their worship. Whether it is an idol or a cow, neither is capable of being a candidate for the sophisticated creator of the heavens and the earth 'the universe'. In other words, people are fully aware that these 'deities' are not capable of creating 'one atom', let alone a universe. As a result, they are scammed into believing these objects are merely a transit to reach the Supreme Creator. In some cases, it is claimed that it is not the objects themselves that are worshipped but rather the spirit residing within them.

Pantheism sells the idea that the universe as a whole is God. Hence, there is no God except that everything within the universe are Gods as such, God is 'everywhere'. Consequently, the act of worship can be directed to everything, even when it is known that these 'worshipped deities' are created beings themselves. This doctrine thrives on the fact that the sophistication of God is evident in many of His creations and creatures. To appreciate the wisdom and power of God, 'a touch of God' is apparent in the existence of His creatures. Just as a touch of Henry Ford is apparent in the cars he created, such as Ford automobiles, or Enzo Ferrari in Ferrari cars, or Steve Jobs in the iPhone, so on and so forth.

An example illustrates this point: the apple tree displays this 'touch of God' every season when it produces its most nutritious apple type, known as the 'red delicious'. This variety packs more nutrients than any other, such as anthocyanidins, polyphenols, and high calcium content. This tree produces such an amazing fruit out of seeds placed under the sand with water a miracle in every sense of the word. Unfortunately, we may not conceive it as such because we take it for granted. In fact, this and similar phenomena are more miraculous than the optical illusions 'magic' we pay to see in theatres, though we perceive it not. However, despite this 'touch of God', the apple tree cannot be nor can it contain even a shred of— the singularity power of God 'Himself.'

To demonstrate this point further: we as humans routinely cut down apple trees. With a right mind, no one believes that when an apple tree is bulldozed, it means God was killed, or part of God was destroyed. Furthermore, anytime a Ford F-150 truck—the best-selling vehicle in the Ford lineup—is scrapped or demolished, it does not mean Henry Ford was hurt. The apple tree is among countless created beings within our universe, including people such as Henry Ford himself. Without exception, they are all the brainchildren of God, just as the Ford Motor Company is Henry Ford's brainchild. Therefore, neither the tree nor any created being is divine, just as much as no Ford vehicle is the human Henry Ford himself.

The Immortal Intelligence Engages with Buddhism

As indicated in our previous discussions on this subject, while the immortal intelligence accepts that 'Buddhism' qualifies 'in one sense' the definition of religion—being 'a way of life', a lifestyle of sorts, if you will, nevertheless, this belief system does not satisfy the requirement of the immortal intelligence that a 'full-blown' religion must provide a clear and reasonable explanation of how this sophisticated universe began, who initiated it, and for what purpose, if any?

The fact that the father of Buddhism himself, Mr Gautama Buddha, said, according to the well-established priori of 'Buddhist Philosophy,' "It is not my business, nor should it be yours, to find out whether there is God...."combined with the fact that Buddha has maintained neutrality in neither affirming nor denying the big question throughout his preaching lifetime is sufficient to disqualify Buddhism from our quest, which is the purpose of our research, as to who qualifies to be the candidate for the 'Singularity' beginning of the universe.

The immortal intelligence holds the position of the founding father Buddha in high regard against anyone else who later, after his death, developed inconsistent theology contradicting Buddha's stance on the subject matter within his doctrine, a rather trendy behaviour protracts which astoundingly is not limited to Gautama

Buddha alone but underpins the experiences of all founding personalities of famous religions without exception from Moses of the Torah, and Jesus of the Gospel even to Muhammad of Islam, albeit to a lesser degree. Why this is the case will be explained later. However, it is apparent that religious clerks, in the name of these admired personalities, have often exploited their popularity to gain power, misguide people, or commercialize these great influential individuals for personal gain. Sometimes, this is done to attract large followings, essentially creating a 'base' or to gratify the organ of power over the land, such as the King, etcetera.

These clerks achieve success by leveraging the 'goodwill-trust' people place in them without scrutinizing their claims against the scriptures. The good news, however, for those of you with immortal intellectual inquisitiveness, is the fact that these founding personalities were clear in their communication in so far as who they were and the message they intended to deliver! For those who genuinely care to know, this clarity is often retained in the scriptural books, even when the organized religious schools would rather you not discover it or attempt to inundate you with additional information to divert your attention. We will engage with this trend further in our discussions, but, rest assured, this phenomenon has occurred more frequently than one might imagine.

While the immortal intelligence dislikes speculation, we will make an exception this time. It is fair to extrapolate that since Gautama Buddha was born and raised within the culture of

Hinduism, he became dissatisfied with its practices of idolatry due to the fact that he had to look into other alternatives. This dissatisfaction mirrors that of young Muhammad in 6th-century Mecca, Arabia. As a result, Buddha had to explore alternatives. It appears that, with all intents and purposes, he defected from his cultural belief system by learning 'Jainism' as a transit stop in the interim before he eventually started preaching what would become 'Dukkha', which is a practical path to eradicating and alleviating poverty and hunger from the masses setting his eyes on 'good causes' by fulfilling what he called the 'noble truths' , which amount to be good life skills. These include what he called the 'noble eightfold path', which encourages individuals to cultivate a good view, thoughts, speech, actions, effort, mindfulness, meditation, and livelihood in order to reach a successful life with minimal suffering, particularly from hunger and poverty. According to Buddha, achieving this benchmark is what he referred to as Nirvana, which he presented as the ultimate prize in Buddhism.

What is astonishingly paradoxical, however, is that despite the fact that Buddha himself never claimed divinity for himself, 'not that the immortal intelligence would ever agree with any human divinity anyway', the Mahayana subsect of Buddhism later elevated him to obtain divine status, without his 'explicit' consent. This repetitive methodology of elevating humans to divine status has also occurred with many individuals who vehemently negated such claims during their lifetimes. A man who refused to entertain answering the big question of who is the likely candidate for the Creator of the heavens and earth could not logically be God the Creator. If Buddha was not divine when alive, how could he suddenly become God after his death?

Consequently, the vision of Buddha has always been a noble cause of alleviating poverty and hunger, teaching sufferers the skills from which to reach 'self-reliance', which shares the thematic benchmarks of UN World Food Programme (WFP) and other international poverty relief agencies, NGOs, and charities. Furthermore, the Samaritan ethics, good morals, and humble human conduct emphasized by Buddha are also core teachings of the Abrahamic religions, without anomalies. However, the Abrahamic belief systems engage with the big question in a more pronounced manner, particularly, their respective central figures, such as Moses, Jesus, and Muhammad, unlike Buddha. This will be the focus of our next topic of discussion.

The Immortal Intelligence Engages with the Hebrew Bible, Moses and Judaism

As soon as you pay attention to this ancient monotheistic religion, you would easily find congruity with the 'Singularity'—the ultimate singular power behind the universe. This is, at this point in our conversation, a well-established fact in both scientific and religious realms, despite some superficial noise to dispute it.

In Judaism, this Singularity 'one' does not somehow become three, and three does not somehow become one, as seen in the doctrinal concept found in modern Christianity famously known as the 'Trinity'. The ultimate God—Ela/Elohim/Allah—is Singular (Echad/Ahad) on His own, without any divisional multiplications, such as the Father also becomes his own Son, and vice versa. The Son is also His own Father, and somehow, this is accompanied by a third sub-entity known as the Holy Ghost that will visit people to convince them of this 'convolutedly tangled' doctrinal concept.

If the above is not mouthful enough, mother Mary is the biological mother of the Son, who is also the Father, and, consequently, she must also be a mother to the Holy Ghost because the three in the Trinity are one, even though somehow they are distinct persons. This means that if you are a mother to one, you must also be a mother to all. Consequently, God is said to have a mother. In this line of thinking, one cannot help but wonder: if God has a mother, where is the father then? Oh yes, the Son is the

Father because they are 'one'—Trinity, remember! Is the mother married to the Son then?; So on and so forth, as this puzzle triggers an infinite floodgate of irreconcilable questions?

To be frank, whether this concept is authentically linked to not just to Jesus but to any of the Abrahamic central figures such as Muhammad or Moses will be scrutinized in later stages of our conversation, particularly in the New Testament section where it is cultivated.

The second thing the immortal intelligence would notice immediately is the name 'Judaism'. Where does it come from? Does it carry any theological significance? Was it part of the teachings of Moses, the central figure of the religion? Was it included in the commandments Moses taught? In other words, was it revealed by Elohim with the backing and authority of Moses?

Unlike the Singularity, this does not pass the test. It shifts the theology from being a concept a 'way to earn eternity' such as observing the commandments, to being associated with particular people/ethnicity and/or race, specifically a tribe for that matter. If someone were to interpret this as 'racial and/or ethnic superiority' it would not be entirely nonsensical. Conversely, in fact, Judah, the man after whom the name is based, is not only one of the twelve tribes of Israelites but the title 'name' also represents an internal rift between Israelites along tribal and racial lines.

Yes, there are internal divisions among all the Abrahamic beliefs, such as Christianity and Islam, respectively. For example, the Schism of 1054 divided the Orthodox East versus Catholic Rome, while in Islam, the Shia broke away from The Sunni, where the latter's split is mainly based on political, ideological, and theological differences. In the former 'Judaism', however, the segregation is primarily predicated on racial/ethnic superiority both among themselves and towards others, such as the Gentiles/Goyim.

After the death of King Solomon, a military general named Jeroboam, from the tribe of Ephraim, allied with other tribes, such as Manasseh, who are the descendants of the Joseph tribe. These tribes deemed themselves to be the most powerful tribes of Israelites. Hence, as such, felt they were rightfully deserving to rule over the Israelites and refused to be governed under the southern kingdom named after the tribe 'Judah', otherwise known as the southern kingdom of Judah. The northern tribes, led by General Jeroboam, went to sectarian war against the southern tribes of Judah as well as Benjamin, who would later be dissolved into the Judaic tribal Kingdom.

The general Jeroboam and his allies emerged victorious, establishing their self-governing kingdom, which they named the Northern Kingdom of Israel, as referenced in 1 Kings, Chapter 12. The immortal intelligence notes that in addition to tribal divisions between the southern and northern tribes 'Rehoboam versus Jeroboam', accusations of polytheistic idolatry practices were levied on both sides in order to score victory over each other justifying their rivalry on merits of religious purity. However, the bottom line

161

aim was to attain power and control over the land at any price, driven by tribal pride.

As it is self-evident in history, the name 'Judaism', which is originated and derived from 'Yehudah', one member of the tribes of ancient Israelites which as a religion identity was founded long after Moses—has no connection to Mosaic theological substance whatsoever. Furthermore, it alienated most of the northern Israelite tribes, such as the Ephraimites, Manassites and others, who made up the majority of the population at the time based on their numbers. These ten tribes significantly outnumbered the two southern tribes. If other Israelite tribes felt excluded and refused such a ethno- racial and tribal-based identity, why would anyone else accept it or feel 'inclusive', using the name of one tribe to represent the entire identity of the religion sends a sectarian and/or discriminative signal to outsiders, even to this day. Evidently, even the great Moses himself would have experienced alienation and exclusion under this racial/ethno identity since his tribal identity which he was born into is Levi not Yehuda/Judah.

For the global population, neither God (Eloh/Elohim/Allah) is a Jew, nor is His requirement for eternal life tied to becoming a member of the tribe of Judah. Since Judah is a race and specific ethnicity, no one can, by elective choice, change their race or ethnicity to become a Jew unless they are born into it. Imagine, with an impartial eye, the legitimacy of a religion named after medieval European tribes such as the Goths, Franks, Angles, Saxons, Slavs, or

Nordics—or the tribe of the Prophet Muhammad, the Quraysh, or even my tribe, the Hawadle of Somalia in East Africa. No one would readily believe such a religion is fairly representing the Creator of the universe—the One who created all people with their different ethnicities, genders, tribes, races etcetera.

With that said, the immortal intelligence draws an analytical comparison with the tribal civil wars that ensued after the death of the Prophet Muhammad. These demonstrate that tribal divisions among Muhammad's early followers were also vibrant. This gave rise to the Umayyad dynasty (661–750), followed by the hostile takeover of the Abbasid dynasty (750–1258), both sub-clans of the larger Quraysh tribe. However, neither of these dynasties were under the illusion that their hegemonic power grabs would stipulate 'lead' to replacing the name of the message brought by Muhammad and the religion of Islam with their tribal dynasties, unlike the Judaic Kingdom.

In other words, no one dared to have the audacity to align the religion's name with their tribal identity. Hence, the Islamic religion remained separate from the secular power struggles of the day, whereas the message left by Moses succumbed to alteration by the secular powers of the day.

However, beyond the name Judah, which should not need to represent the message of Moses, there are striking similarities with the message of Muhammad of Islam. So much so that the immortal intelligence struggles to comprehend how one can affirm one and reject the other—that is Moses against Muhammad. How can an authentically sincere impartial soul embrace Kosher but deny Halal? How can an impartial sincere soul enroll Halakha but reject Sharia unless it is done so through the lenses of partial eye rather than an impartial eye?

The veracious common route between these religions lies in the same God (Eloh and Allah), the same father, Abraham, with his sons Ishmael and Isaac, and the same core requirement of singular worship directed towards the singular God, Eloh/Allah, alone. Without any preconditions, upholding His sovereignty has always been the sole path to eternity. Ask your spouse, "What is the one thing you think you deserve from me?" Chances are, they would reply "Unconditional love". Is not the Creator more deserving of that love than your spouse?

The immortal intelligence must be fair and acknowledge the existence of an access for those who want to convert to Judaism, even after you somehow convince your immortal intelligence to get over it with 'grain of salt' your new identity would be someone else's racial and tribal ethnicity, You would face a pretty steep uphill climb with no guarantees at the end whether you would be accepted to become Jewish by conversion. The process speaks for itself and readily accessible for anyone who is interested, one thing that stands out beyond the ritual undertakings is you need to express your wish to convert to Rabbi, who 'in the tradition' must reject three times to see whether you are sincere or not! The converting person is astonished by the prospect of facing human acceptance in God's religion and perhaps is wondering am I here to access people's wealth, applying for a loan from the bank or God the Singularity of the universe?

How is it possible anyone but God to know what is in my heart due to genuine sincerity is internal 'state of being' away from the eyes and recognition of peers? After one goes through at least a year-long process, you would end with a certificate that says you are 'good to go', but there are no guarantees if every denomination would accept you are naturalized Jewish person thereafter. The immortal intelligence questions why does one need someone else's approval to be servant of God? Should not God be accessible to all equally?

Would you reject Jesus for promoting equality—that all men are created equal under God—and recognizing that you cannot attain equality until you give others the rights you claim for yourself?

This is referenced in Matthew 7:12 in the New American Standard Bible: "In everything, therefore, treat people the same way you want them to treat you, for this is the Law of the Prophets". Would you reject Jesus for sharing that the merciful God would extend His grace to all peoples, regardless of lineage, ethnicity, race, or color, including the Ishmaelite Arabs? This is referenced in Genesis 21:18: "I will make him [Ishmael] into a great nation".

Jesus, doubling down on this, stated in Matthew 21:43 that the Kingdom of God is not a 'blank cheque' monopoly belonging to one race/ethnicity with impunity—unlike the disgraceful U. S. veto in the Security Council (SC) which was hired against the humanitarian world of 120 nations who proposed no less than 5 Resolutions to stop the senseless suffering of the people of Gaza as of November 2024. Instead, it can be taken away and distributed to others who would bear its fruits authentically: "Therefore, I tell you that the kingdom of God will be taken away from you and given to a people who will produce its fruit".

Would the above serve as a theological or moral ground to reject Jesus? As referenced in Matthew 21:46, the high priests of the temple in Jerusalem—the Pharisees—plotted to silence him, but they knew the general public had already accepted him as their 'Prophet'. However, they feared the ramifications in doing so may cause as the people 'the crowd' held that he was a prophet: "They looked for a way to silence him, but they were afraid of the crowd because the people held that he was a prophet".

Would you reject Muhammad just because he is from a different ethnicity, a cousin, an Ishmaelite? No wonder he is referred to as the 'rejected stone' in the Bible, as seen in Genesis 21:13: "And of the son of the slave woman I will make a nation also because he is your descendant" . Would you reject him because his mother was an Egyptian slave, Hagar, as referenced in Genesis 16:15: "And Hagar bore Abraham a son, and Abraham gave the name Ishmael to the son she had borne"?

Is it because the coercive conformity of Judaic hegemonic power was exerted upon the other eleven tribes, as seen earlier? Or is it because you live in a pretentious and forgetful state of mind, ignoring the fact that you Israelites, including Jews, all without exception incessantly remained Egyptian slaves yourselves—the 'house of bondage'—with no ability to resist? How can you ('Judaism') claim ethnic superiority over Ishmaelite Arabs because his mother was an Egyptian slave when all Israelites themselves remained Egyptian slaves for 400 years? It is illogical stance! You gained freedom only through the Merciful Eloh/Allah, who had no obligatory requirement whatsoever 'single-handedly' out of His mere mercy freed you, was it not through His compassion, via Moses, that you gained pragmatic freedom? This is referenced in Exodus 13:3, as discussed earlier.

How do you justify to your moral hearts the redemption of brutally killing all the historic prophets, such as Zachariah and John the Baptist? Jesus hoped for you, yet you rejected him. As referenced in Matthew 23:37, Jesus expressed a wish for your redemption, likened to the height of motherly love as a mother gives to her children, but you rejected him out of stubbornness, pride or misunderstanding: "O Jerusalem, Jerusalem, the one who kills the prophets and stones those who are sent to her! How often I wanted to gather your children together, as a hen gathers her chicks under her wings, but you were not willing. "

How do you justify to your moral hearts the destruction of Allah's house—the Al-Aqsa Mosque, with its golden dome—the same God of Abraham and Moses by creating and supporting global lobbyist groups such as the Temple Institute? These efforts aim to replace it with a symbol of Judaic hegemonic power, claiming it as the house of the same God, Elohim. How do you alienate Allah while justifying love for Eloh when they are the same entity?

The immortal intelligence questions the rationale of expecting the true followers of Jesus to support the very same subject that you rejected in Jesus himself when he hoped for your forgiveness from the Lord.

If God were to have created me from Jewish lineage, that is, If I were a genuine Israelite descent from the innocent common population, I would sincerely be baffled at the prospect of a third party who has no 'any' apparent direct stake to emphatically declare to me I would build a house for you even If you did not want to do it for yourself. Such statements are often coded from the mouths of non-Jewish Zionists, "Even if there were no State of Israel, we would create one", for what 'intent and interest' it serves to them for such a colossal undertaking? Do they love me that much in the sight of the Singular Elohim, the ultimate target according to Moses, I would ask myself?

Further, why the Fascist far-right Hitler sent his top diplomat Joachim von Ribbentrop to negotiate with the British government who took over the Holyland from the Ottoman Empire during the

168

First World War 'the great war', what was known as Palestine at the time to settle the Jewish communities of Europe, although predominantly from Germany at this stage, via shipping them off to the middle east under the collaboration agreement between Zionists and Nazis known as The Haavara Agreement (1933). Do not be astounded by the existence of such agreements in the past. After the negotiations collapsed and subsequently, his sinister genocidal ethnic cleansing ensued before and during the Second World War, had the tyrant not unilaterally started the war or lost it, what would have been the fate of the Jewish communities?

Why many Muslim leaders put their lives at risk, such as the Imam of the Grand Mosque of Paris, Si Kaddour Benghabrit, who hid two hundred families of Jewish descendants during the Nazi-occupied France who faced certain genocide and destruction? Further, why many learned scholar Rabbis at the time, such as Joel Teitelbaum and others, opposed Jewish settlement in the Holyland to be achieved through the suffering and bloodshed of others 'Abrahamic children?'

While these questions are food for thought, I myself do not have all the answers relating to them. Nevertheless, the afore incidents are thought-provoking facts that took place in history. It was Winston Churchill who said you need with a critical eye to look into history as far behind as possible to see as far ahead as possible: "The farther back you can look, the farther forward you are likely to see".

Did the Artificial Orbit liquidate your intelligence that you calculate the Creator of the universe 'somehow' benefits from or must abide by your personal opinion? He is the One who has already decreed our time of death, regardless of who we are, without our consent. Whatever title is vested in you or whatever weapons your army amasses, these human shields amount to nothing against His absolute will—the decree of God. You already know this if you read the scriptures.

This means: do we expect God to accept us, or is it the other way around—that we must accept Him? The concept of 'subjective acceptance' of God, rather than the objective acceptance of God, deserves reflection.

Many people historically have fallen victim to the concept of 'subjective acceptance' of God, also referred to as 'conditional worship' or 'transactional worship' of God. This is when your acceptance of God is limited to the extent that you are receiving what you expect from Him. If the provisions of God differ from or contradict your expectations, you may reject Him, or alternatively, pretend to be a follower of God 'hypocritical worship', or even change the message itself until it becomes palatable to your liking. Hence, in this mode, you subconsciously want God to compromise with you in exchange for your acceptance of Him.

This is why many people go against God, not because they have verified His existence or lack thereof, but based on their opinion disagreements with His decisions or things out of their control happening to or around them rather than verifying the big question: is there an entity behind the universe transcendent to universal natural laws? An example is conferred: suppose someone loves drinking intoxicants and says . "I cannot imagine accepting a religion where drinking alcohol is prohibited". This has nothing to do with verifying whether there is a God or not. Imagine you do not agree with some policies of your government—that disagreement does not negate the fact that the government is the lawful authority of the land. Similarly, many reject God because of tragic incidents or suffering: questions like, "Why are children or innocent people suffering?" are common examples.

For some, their 'subjective acceptance' of God is conditioned on their perception of justice. For instance, "I cannot follow a religion where men are allowed to have more wives (polygamy)" or similar concepts in the same line of thinking, such as a religion where its justice system can confer judgments of capital punishment or cutting off hands and the like. However, the thing we should disagree with the most is the concept of our 'death' itself. Yet simultaneously, death is the most enforced law of God—without any exceptions, no ifs or buts. We are all facing to taste death, irrespective of status, title, power, or wealth. Notice that none of these objections address whether God exists?

The immortal intelligence, however, grants the appreciation that the Almighty God is unlike humans, who are prone to making errors in judgment. God is measured by a higher standard, as a result, in order to understand the occurrence of inexplicable scenarios—such as why God allows the death of young people or the suffering of innocents, why God allowed the prophets to be killed in Jerusalem, why John the Baptist was beheaded and his head presented on a platter (referenced in Matthew 14:3-5), or why Zechariah was stoned to death in the courtyard of Jerusalem's Temple while preaching worship for the Singular God alone; no idol reverence (referenced in 2 Chronicles 24:20)—one must consider this: the people of understanding say, "This happened by wisdom only known to the Almighty God". Because, after all, it should not have happened as far as our limited brains are concerned.

Otherwise, there are only two outcomes from this: either there is a very sinister God allowing His good servants to be harmed, or there is no God at all. Therefore, we humans are naturally evil and have the propensity to cause bloodshed and corruption.

To respond to the above, an example is conferred: Imagine a university student complains to his/her professor, "Why do you give us assessments, essays, and assignments? Why do we need to spend hours in the library, hours and hours in perpetuity, stressed late night studying during exam month, suffering depression and anxiety as well?" The professor responds these sacrifices and sufferings achieve two goals: first, to differentiate which one of you worked harder over the other so that it would reflect on his/her transcript and degree certificate; and second, through these struggles, sufferings and pain you would achieve and gain a successful life in the real world after graduation.

It is an oxymoronic predicament for those who blame God for the trials in their lives often give a standing ovation to Winston Churchill's famous speech on the eve of the Second World War to the British parliament and the public: "I have nothing to offer but blood, toil, tears, and sweat. We have before us an ordeal of the most grievous kind...." In other words, prepare yourselves for some turbulence to achieve lasting security later.

However, yet when the Singular God says in Koran 2:155-157 Prepare yourselves tests and trials will be coming on your way, It 'suddenly' is unreasonable to the very same people who applauded for Winston Chrchil "*We will certainly test you with a touch of fear and famine and loss of property, life, and crops. Give good news to those who patiently endure, who say, when struck by a disaster, 'Surely to Allah we belong and to Him we will ˈallˈ return. ' They are the ones who will receive Allah's blessings and mercy. And it is they who are ˈrightlyˈguided.* May God save us amen.

That said, the above disagreements more often come from lay people like you and me. However, some of the scholars also have their equivalent to that end but variant, their own form of 'subjective acceptance', which similarly amounts to a rejection of God. In their case, however, they have more definite knowledge of God 'Allah/Elohim/Ella' unlike the layperson described above and yet

173

they disagree with some of his Divine policies, often concerning personal power and/or prestige-related appointments.

For instance, many of these scholars 'historically' expected and thought that they deserved to be upgraded to be included with the cabinet of prophets of God due to their physical dedication in relentless worship of God. When this promotion was not coming through, hence some type of rejection or modification of the text would ensue or, alternatively, some of them wanted to keep an 'embargo' on the mercy of God conferred on others—similar to some of the world powers of today carrying out an 'embargo' on the smaller States, usually the rival tribes that they wish to never experience the mercy of God, by that, there should not be a prophet risen from 'the other side' for instance, and the like.

The one who has fallen the most victim to this concept is the entity known as 'Iblis' or 'Lucifer' in the scriptures. He would have remained an angel had he not infringed on the universality of the 'mercy of God'. More importantly, his infringement violated 'the sovereignty of God'. The failure on his part was his inability to comprehend that all blessings belong to God, and it is within God's legitimate right to distribute what is His as He pleases. The masculine word "He" should not be understood in our day-to-day vernacular but as referring to an entity transcendent of gender variations, much like when we say "mankind."

The scriptural story of Lucifer (Iblis) envying Adam for a blessing he received from his Lord is remarked in all the Abrahamic scriptures with more or less similar narrative (referenced in Genesis 3:1-24 among many other books within the Bible, the immortal intelligence will refer the one in the Koran for its proximity to its central figure Muhammad which is compiled less than two years after his death, by that it means, the Old Testament was compiled over a thousand years after the death of Moses 'its central figure' which we will engage with later in our conversation, while the manuscripts of the New Testament are not even written in Aramaic, the language that Jesus had spoken, not to mention the fact that it went through multi-layered linguistic translations along with 'unknown' clergy interpretations. It has been translated from Greek, its oldest manuscript, anything older than that has been lost, from Greek to Latin, and from there, it has been translated to English and other world languages, including Hebrew and the like; yes, even the Hebrew was translated from the Koine Greek Manuscripts because that is the oldest manuscript for the Bible. This is further compounded by the lack of consensus with regard to what books form or constitute the Bible of today. The Vatican Catholic Church approves 73 Books, whilst the Protestant church approves 66 Books, not to mention other branches of Christianity, such as the Orthodox and the excommunicated branches, such as Nestorians and Unitarians who were known to be early followers of Jesus the Messiah.

Subsequently, the Koran tells us Iblis ('Lucifer'), as part of the Jinn creation of Allah (referenced in Koran 18:50), was included within the angelic ranks, despite his creation being a variant of energy, 'smokeless fire' to that of the Angelic beings 'light' (referenced in Koran 55:15). The fact that he was part of the angels means not only did he have an in-depth knowledge of God, to the contrary of the laypeople discussed above, but he supremely worshipped Him too. Hence, his attainment to be qualified to take angelic status thus taken in with the ranks of Angels, on face value, Iblis would reign supreme over Angels due to his 'freewill' status, which angels do not possess. Consequently, his worship of God carries more weight than the angelic beings, without any freewill choice or without any power to disobey God, much like a software implementing the coding effect by its programmer or when someone presses the tap button call on his 'Mobile' and as consequence, the phone calls the intended person. No one is thrilled to say the phone was sincere to me, it fulfilled my expectations from it, because it lacks freewill. Such are the angelic beings, as the scriptures inform us they do not possess the freewill to reject God.

Prior to the creation of Adam, thus Mankind with highly 'intellectual' profile status on the planet Earth, Allah would announce to all the Angels, including Iblis ('Lucifer'), the coming of human beings with honorable status conferred upon them, who would take the role of 'Caliph ', meaning the one who is in charge of the planet Earth (referenced in Koran 2:30). Therefore, the requirement to give respect and esteem to Adam was placed upon the Angels as soon as the decree of Adam's existence would be put into force in the form of bowing down 'prostration' offered to Adam by the Angels with no preclusions, which would become a prohibited 'act' after the coming of Muhammad—'no prostration shall be offered to any other than the God Allah'—because in the old Abrahamic traditions while bowing to other than Eloh for worship purposes was prohibited, however, carrying out the act out of respect was allowed.

Therefore, if you find yourself to be abhorrent of the title 'Caliph', you may be an inadvertent emulator of 'Lucifer' according to the Koran, perhaps without the cognition of its historical reference. The other interesting contravention to the status of mankind being 'Caliph' in charge was other Angels who showed a degree of concern for world peace due to Mankind's propensity to violence, abuse of human rights and cause offensive wars while possessing supreme power of weapons of mass destruction with an inclination for hegemonic suppression of one another, referenced in the latter part of the same verse as emphatically detailed by Professor Jeffery Lang in his famous lecture of "The purpose of life" available on YouTube.

The decree was put into force, and Adam came into being as he quickly showed a phenomenal capability and trait of intellect comprising both the moral and learning capabilities of mankind, which sets him apart (referenced in Koran 2:31-33). These above capabilities were sufficient to address these concerns, however, Iblis

177

rejected the requirement to give Adam honor and esteem based on what we would regard as racism today in our lingua franca. Why he rejected when asked by God, his response was, "I am created from fire and in contrast, Adam is created from 'dirt' clay, therefore, I am superior to him" (referenced in Koran 7:12), rather Adam should show me esteem and honor instead is inferred from his comments.

What other definition befits 'racism' except the arrogance to hold your natural traits bestowed on you by God to make yourself superior to others, such as "I am white, therefore I am superior to the black/brown fellow human beings" or "I am a Jew, therefore, I am superior to others 'Gentiles/Goyim' i. e. non-Jewish race on the basis of my ethnicity, " or similarly, "I am an Arab, therefore, I am superior to others 'Ajams' i. e. non-Arab race on the basis of my ethnicity" and the like.

Another comprehension failure or grave mistake on the part of Iblis ('Lucifer') was that superiority is effort-based, not creation-based. An example is conferred: a good Jinn who does the works of God should stand superior before God Allah than a bad tyrant human who rejects his Lord Allah. Iblis ('Lucifer') Due to acting arrogantly before God, he was expelled from the presence of the divine (referenced in 7:13).

In Koran 17:61, Lucifer vowed to Allah that these human beings, whom you have 'supposedly' honoured over me by making them 'Caliphs' in charge of the planet Earth, would not repay the favor to you. "If you grant me a respite of long life, I would see to it that they are unthankful to you." In 17:65, Allah assured him that Iblis would not have any 'influential' power over the 'sincere' servants of God—very interesting verse.

In Koran 18:50, Allah reminds mankind that *"I banished Iblis from my paradise 'Kingdom' because he refused to show honor and respect to Adam 'your father' then will you take him and his family as allies and friends in lieu of me, while he and his family are an enemy to you. What a treacherous treatment you return to me!"* In 18:51, Allah further clarifies to what extent powers or lack thereof were given to Iblis. As Allah says, *"I omitted them to witness any of the creation of the heavens and earth including how I created themselves, nor anyone who intents to hurt others is my allies."* In other words, it is inferred that Lucifer 'Iblis' is not capable of creating a life of any sort from a small fly to a beast and anything in-between, nor anyone whose intention is to disgrace others would be a member of Allah's camp, another interesting verse.

The immortal intelligence derives from above that Adam and human beings were given the high responsibility of being 'in charge' of the planet Earth due to the high intellectual aptitude and capacity that was bestowed onto them. Therefore the declared war of Lucifer to Adam and his descendants before Allah would be occurring predominantly in the 'intellectual sphere' that is because in order to demonstrate you are smarter than someone else, hence, you are superior to them intellectually, the highest level of achieving that is to 'outwit' them, to 'deceive', to scam them, to trick them, to mentally humiliate them, disgrace them etcetera.

To muddle-up the good and the bad and swap the right with the obvious wrong to the extent that a teenager is encouraged to change his/her natural gender with irreversible ramifications without the consent of his/her parents but cannot buy a cigarette from the shop due to its health consequences, it is also why some people like to play 'poker' games in gambling so that they feel smarter or have the illusion to posses higher IQ than their counterparts whom they are outwitting and/or 'scamming' their money from, perhaps it fills-up a gap 'ego deficit' in their hearts.

What the scriptures, the Koran to be specific, is denoting is the fact that 'Satanism' is rather a 'conceptual act' rather than a natural being contrary to common misunderstanding, therefore there is nothing inherently natural about it, in other words, nobody is born to be Satan or evil for that matter as seen with Lucifer being dedicated worshiper of God to the level of Angelic status, a scholar 'in his own right' in terms of his direct knowledge of God being substantial by any measure, never the less, on the other side of the spectrum, even being a great worshiper or Scholar would not shield you from falling into this failure, becoming an angel 'in and of itself' would not serve immunity from falling into this grievous miscalculation about the Singular Creator of the universe.

The good news, however, is, as the Koran clearly states, with every mistake, there is an opportunity for reformation both Iblis

'Lucifer' and Adam were offered the opportunity to redeem themselves because they both made mistakes. Adam ate the forbidden tree while Adam and Hawa'Eve' took the opportunity to repent, referenced in Koran 7:23:*"Our Lord! We have wronged ourselves. If You do not forgive us and have mercy on us, we will certainly be losers".* Iblis 'Lucifer' on the other hand, has turned down the offer 'to date' to rehabilitate himself, referenced in Koran 15:39-40 *"O my Lord! Because You misled me, I shall indeed adorn the path of error for them (mankind) on the earth, and I shall mislead them all, except your impartial, honest, sincere servants from amongst them".*

The consequence of one's action is either for his benefit or detriment; hence, there is nothing to transfer of responsibility or burden of sin to someone else. No original sin is to be transferred whatever, Adam's sin and downfall is for him, so as Eve and so as Iblis. This is as referenced jointly in the Bible as well as the Koran, in Ezekiel 18:20 "The soul who sins shall die. The son shall not bear the guilt of the father, nor the father bear the guilt of the son. The righteousness of the righteous shall be upon himself, and the wickedness of the wicked shall be upon himself". As well as the Koran 2:41 *"This is a people that have passed away; they shall have what they earned, and you shall have what you earn, and you shall not be called upon to answer for what they did".*

Further, the Koran 38:18 would often repeat no person will take the sin of another, even if it be a loved one or relative *"And no*

181

bearer of burdens will bear the burden of another. And if a heavily laden soul calls [another] to [carry some of] its load, nothing of it will be carried, even if he should be a close relative. You can only warn those who fear their Lord unseen and have established prayer. And whoever purifies himself only purifies himself for [the benefit of] his soul. And to Allah is the [final] destination".

The immortal intelligence points to another interesting highlight: the Koranic version of Genesis in the subject matter differs from that of the Bible, as the Koran calls itself 'the guardian book' over the original messages revealed to Jesus and Moses respectively 'Injiil' Gospel and Torah referenced in Koran 5:48 *" We have revealed a book........that is guardian upon the scriptures before it......."*, as it accuses alterations occurred on their messages over time referenced in Koran 2:79 *"Woow to those who ascribe 'the scriptures' from man's opinions in order to gain 'small' profit and say this is from God......."*

Eve 'Hawa' in the Koran is not discriminated against because of her gender, she does not have increased pain with child-birth due to God's curse on her for giving Adam a fruit from the forbidden Tree, thus remaining inferior to man as a result referenced in Genesis 3:16, conversely, mothers stand higher honor over fathers, three folds over to be specific, referenced in authentic Hadith Bukhari 5971, and one of these folds is due to the pain the mother undergoes during the process of childbirth which the father does not, quite the opposite, is it not?

182

The Koran emphasizes both Adam and Hawa 'Eve' were intellectual beings who made a mistake equally, co-sharing the fault evenly, if anything, it appears Adam is more culpable of the two due to the fact that the Koran subjectively informs us that Adam disobeyed his lord and ate the tree referenced in Koran 20:121 "*So they both ate from the tree and then their nakedness was exposed to them, prompting them to cover themselves with leaves from Paradise. So Adam disobeyed his Lord, and ˈsoˈ lost his way*".

Further, if you ever wondered that Adam is blameworthy for mankind receiving expulsion from paradise, you are not alone my friend, even the great Moses held the same opinion according to the Islamic Hadeeth literature, where Moses blamed Adam for getting 'us' expelled us out of paradise, but Moses did not win in this argument because our father Adam reminded him we were always going to be tried and tested on planet Earth with our freewill and high intellectual capability, in other words, it was always our destiny, as Allah says in the Koran prior to creating mankind Koran 2:30"......I will establish on earth human *caliphs* 'an authoritative entity'......" If you are interested in this fascinating moral argument between these great influential giants of the highest magnitude, see 'authentic hadith' Hadith Sahih Muslim 2652a.

This book is not precisely the story of Adam and Iblis 'Lucifer', or what is Satanic versus what is Angelic, but to research who deserves to be crowned the creator of the heavens and the earth via utilizing our immortal intelligence, thus, we resume our engagement with the Bible and Judaism.

As earlier mentioned, the Old Testament—also known as the Hebrew Bible or the 'Pentateuch' comprises five books: Genesis, Exodus, Leviticus, Numbers, and Deuteronomy. These books had a serious time gap between them and their central figure, Moses, let alone their proximity to Moses himself. In fact, it was proven that these books were not written by Moses, thus setting aside the claim that they have one author. Even the Rabbinical Talmud, referenced in *BabbaBatra 15a*, which is most inclined towards Mosaic authorship of the Hebrew Bible, which itself was written 300-500 years after Christ (AC), admit to themselves the existence of very serious doubt whether Moses wrote parts of the Bible. In Deuteronomy 34:5–12, where the detailed funeral of Moses is narrated by a third party, Rabbi Shimmon or Nahhima(depending on which version), shares with his friend Rabi Yahude 'Juda' his suspicion of the third party may very well have been Joshua since he was his trusted student prophet.

Bear in mind that Moses preceded Jesus by at least 12 to 14 centuries. However, the sure thing is that no one knows who could have authored those verses. Then the natural question arises: If it is proven that Moses did not write the entire Hebrew Bible, then who did, and why are there so many contradictions?

Bible scholars, including Professor Joel S. Baden in his work 'The Composition of 'Pentateuch', which literally means the composition of the Hebrew Bible, agree that the first written work from Moses was produced within the 1st century before Christ (BC). This means, if you do the math, it was at least 13 to 14 centuries after the death of Moses. Imagine if the first time a written Koran was produced 14 centuries after the death of Muhammad, which is 'almost' in the 21st century.

The disparity of narrative contradictions is too numerous to enlist them all here, however, Professor Baden and others, such as Professor Ehrman, used these differences as evidence to deliver the undeniable fact which proves that these books were authored not only by multiple persons but evidently by conflicting writers and ascribes at that, due to the absence of coherency in conjunction with blatant discrepancies, from simple details to major historical events and incidents that have changed the course of the Israelites' era. For instance, the father-in-law of Moses is called Reuel in Exodus 2:18 or Jethro in Exodus 3:1 to which mountain did God speak to Moses? Was it Mount Sinai (Exodus 19:11) or Horeb (Exodus 3:1 or Deuteronomy 1:6)? Where did Aaron die—a significant incident in the tradition? Take your pick: he dies on Mount Hor (Numbers 20:23-29)or on Mount Moserah (Deuteronomy 10:6). The theme of mismatching sentences and events about the same subject matter is consistently the same and abundant throughout the five books.

Furthermore, Professor Ehrman, in his 'which is it?' hallmark infamous lecture keynote speech, points out, based on his book The Bible: A Historical and Literary Introduction' in Exodus, God (Elohim) performs ten plagues through Moses to exert drought

185

pressure on Pharaoh in order to coerce him into releasing the Israeli slaves. In the fifth plague of the ten miracle performances, referenced in Exodus 9:6, the Lord kills all the livestock of Egypt. However, just a few verses later, in plague number seven, which Moses performs, He kills yet again all the livestock of the Egyptians in (Exodus 9:29-30), which were supposedly all killed in plague number five. Hence, the infamous 'which is it?' objection raised by Professor Ehrman.

However, the Immortal Intelligence must view the above discrepancies for what they are: minor memory inaccuracies except for two things. For one, these unknown narrators and scribes of the Bible making such mistakes, which proves uncertainty as to the authorship of the Bible. And two, the Bible cannot be the ultimate word of God 'verbatim', otherwise, there should not have been readily available contradictions. Above all, astoundingly, what Professors Baden, Ehrman, and others point out in the 21st century is yet an implicit affirmation of the Koranic account on the matter, asserted in the 6th century, that scribes have altered the word of God, injecting human opinions and doctrinal views, as the Koran says yet another verse, Koran 3:78 which, reads part of them 'scholarly clergy-man' deliberately distort the book and pin on God statements that does not belong to God: *"There are some among them who distort the Book with their tongues to make you think this ˹distortion˺ is from the Book—but it is not what the Book says. They say, 'It is from Allah"——but it is not from Allah. And ˹so˺ they attribute lies to Allah knowingly.* "

This verse primarily applies not to the secondary evidence questioned above but to the major primary contradictions that clearly show there have been human interventions into the Bible. These interventions include primary scientific errors within the Bible that demonstrate 'the Singularity' power behind the universe did not verbatim author the Bible. Rather, primitive, unknown human civilizations of the past mixed their personal views along with God's speech. This is the reason why 'they' changed their claim of the direct word of God to the inspired word of God, as if God would inspire colossal mathematical and scientific errors in the books of the Bible would make it 'okay' acceptable to human intellects. The immortal intelligence will further engage with this during the New Testament, 'the Gospels', and Jesus Christ next.

The Immortal Intelligence Engages with The New Testament, Christianity And Jesus Christ

Metaphorically speaking, to many in the Christian background world, it is likely that 'at some point' while you were just an innocent kid, your parents took you to a church of whatever denomination. Whether it was the Protestant divisions, such as Methodism, Baptism, Pentecostalism, or self-reformed evangelicalism, amongst others, or Catholic divisions, such as Roman Catholic or the breakaway churches of Apostles by your 'state' country, such as the Brazilian Catholic Church, the Philippine Independent Church etcetera or Orthodox divisions, via 'state' yet again, like the Russian Orthodox Church, Greek Orthodox Church, or Egyptian Coptic Orthodox Church, inter alia others. Chances are you heard the priest proclaiming that you are 'all set' for a paradise destination. Why? Because Jesus Christ, the only 'begotten' son of God, died for the sins of humanity, thereby making us 'supposedly' sinless beings on the planet Earth, irrespective of our actions. This is an attractive proposal by any measure due to the no-effort requirement attached to the deal, basically offering 'impunity', which is why people would love to join these fancy and powerful organizations which operate beyond the scope of 'accountability of justice'. Whether it is having limitless 'veto' power in the Security Council of the UN or becoming a 'made man' in the Mafia, 'Cosa Nostra', with the structural power attributed to Charles 'Lucky' Luciano in the

1930s, the net result is the same, meaning in their relative worlds: 'You can do anything without serious consequences'. You are protected, hence 'impunity' of sorts, as they call themselves the bosses, the 'Untouchables'.

The process became known as 'Atonement'. However, what you may not hear is that the Atonement doctrine 'in and of itself' is not agreed upon, and why Atonement was needed in the first place is the central curiosity! In other words, why did Jesus, or anybody else, needed to die for humanity? The answer is as shocking to the oblivions as it is demeaning to the Creator Singularity 'Elohim/Ella' the Almighty. This is also a subterranean doctrine under Atonement, which is dressed up as the 'Ransom Doctrine or Theory', as it provides answers to the basis of why Atonement was needed in the first place. Its roots go back to the story of Adam and Eve in Genesis, where the Bible version tells us that Eve enticed Adam to eat from the forbidden tree. As they both ate from it, they committed the 'original sin'. Hence, at that point, Adam and Eve, along with their progeny (humanity), entered 'in the hands' under the control of the devil. Therefore, God the Almighty was required to pay the devil 'a ransom'. Hence, He sacrificed His only son in order to satisfy the ransom payment to the devil for the salvation of humanity.

Which, 'by default', dissolves the Almighty's power as God in this discourse. Any soul with an ounce of purity can see that a God required to pay a ransom to the devil with His only son has 'tones of explaining' to do as to justifying His title in order to remain God, to say the least! Consequently, other theories were developed which problematized the above such as Christus Victor, which subtracted the fray from the devil. In other words, Jesus Christ's death was said to fulfill the purpose of demolishing what it called 'evil', hence defining 'evil' as three things overarching components : 'sin, death, and the devil'. This doctrine utterly fails according to its set objective. Jesus has 'died', according to the Bible, while the three things that were supposed to be defeated are still lurking abundantly, with COVID-19 and other worldly disasters, including man-made wars, increasing the death rate. Sinning has also become a trendy wealth generator. Perhaps the poor utopic Bishop Gustav Aulen, the author of this theory, has never vacationed in 'SinCity' and played a million-dollar roulette while, in the midst of the adrenaline, lost the future of his family. Or perhaps he has not checked the themes that sell highest in Hollywood and the entertainment industry as a whole: 'drugs, violence, and sex'.

What is astonishing is that Las Vegas, the 'Sin-city', as a resort city for holidays, naturally does not own any of the holiday qualities or features, such as beaches, great landscapes and natural palm trees, that your family children are typically looking forward to in a holiday destination. It is as if, mainly, all of the amenities featured in this Sin-city are sins, according to the Biblical accounts. The menu offers a variety of gambling with diversified brain intoxicants, including recreational drugs and the like, with a spectrum of prostitution to top it off with the main course. Hence, one would struggle to have any other intention for travelling to 'the Sin-city' except to sin—apart from the occasional boxing matches, where you are baited into doing the above anyway.

Furthermore, did the Bishop make the assumption that somehow the devil retired two thousand years ago at the 'crucifixion' of Jesus, as claimed in the Bible? These 'drugs, sex, and violence' thematic trends are pointing exponentially upward in every matrix and chart in the graph of sales. Not to mention all the religious fanaticism that has killed so many, together with rampant gun culture. 'Children' and teenagers with machine guns shooting randomly in school hallways and classrooms alike, without any discrimination, reservation, or reason.

Perhaps, as a child with your family, you heard the pastor exclaim infamous Bible codes like John 3:16: "For God so loved the world that He gave His only begotten Son, that whoever believes in Him should not perish but have everlasting life", along with its sister code, John 1:1-14, which reads: "In the beginning was the Word, and the Word was with God, and the Word was God...and the Word became flesh and dwelt amongst us...." Certainly, these pastors hope that you would never activate your immortal intelligence and inspect the claimed assertions, which actually defeats why do you have intelligence at all. In fact, why would you even claim you are a human being if you will not inspect what is on offer in the menu?

Another 'elephant' dispute standing out in the Bible is the difference of who is to be God, the 'Singularity' Almighty, between Jesus Christ 'himself' on one side, teaching the Singular unitary God represented in all the 'other' Abrahamic scriptures such as the Hebrew Bible and the Koran, versus the organized churches of today on the other side, preaching to the masses the manmade 'doctrine of the Trinity' on the back of an invented free ride called 'Under-Grace' in order to bait the masses. Yes, you read that right— Jesus versus the Church. Despite many may not see this as a scientific incompatibility, the immortal intelligence brought this issue herewith due to its magnitude and enormity.

Henceforth, despite the outward denominational variations of these churches, nevertheless, their core theology 'belief' goes back to what is known as 'The Nicene Creed'. Do not be confused with 'The Apostles' Creed'. It is basically the same doctrine dressed up with a different package in that they both intend to make Jesus God the Almighty. Therefore, what is this creed? It is the prerequisite requirement that you have to believe that Jesus is Divine 'God the Almighty' in order to qualify to be Christian. This is what they mean when they say the doctrine is canonized, ousting all other strands of the followers of Jesus—mainly what was known as Unitarian Nazarene followers of Jesus at the time, who believed Jesus is a prophet of God, referenced in Matthew 21:11: "The crowds answered, "This is Jesus, the prophet from Nazareth in Galilee".

Consequently, Christianity, which ought to have been learning and upholding Jesus' message, has been moved away from that to a 'new set of beliefs' which entails the belief in the Trinity being a compulsory 'incoherent and convoluted invention'. This modified and increased the 'Unitary' God to a 'Trinitarian' God, via upgrading Jesus without his consent to be God or the begotten Son of God, which essentially means the same thing in so far as human father and his begotten son are essentially another human. Therefore, you would end up with two humans who are the same species, which also means humanity 'itself' as a species merges with the God—the one who caused the 'Big Bang', who was always separate from the human species historically.

Before we uncover the fact that this doctrine is not revealed by God (historical fact) but is man-made and developed 'cooked' many years and centuries, through many stages, let us try to understand what the Trinity is and, further, how Jesus himself disagrees with this doctrine within the Bible—the very doctrine that is 'supposed to' promote him as a Divine God. What an oxymoronic stance of Jesus against the Trinity doctrine.

All the churches of Christianity today, notwithstanding their nominal divergences, without any exclusion, teach this doctrine of the Trinity, which was canonized 'enforced' in the Council of Nicaea by the Roman Empire in 325 AD. Notice, over three hundred years after Jesus, it was based on the only 'direct' Trinity verse in the Bible, which consolidates the three distinct personalities who are yet somehow one person in the Bible: 1 John 5:7: "For there are three that bear record in heaven, the Father, the Word, and the Holy Ghost: and these three are one." There is no other verse in the Bible where the three are synonymously mentioned side by side as the one 'Trinity'. Otherwise, the three entities are mentioned throughout the Bible as God the Father ('Eloh/Allah'), the Holy Spirit, and Jesus Christ as separate entities. Notice that even in this verse, one has to interpret Jesus as the Word because Jesus is not mentioned verbatim.

Therefore, the Trinity, as taught by the Church, is the combination of three persons: the Father 'Elohim/Ela/Allah', Jesus the Son as well as the Word, and the Holy Spirit, who is yet still another third entity. Consequently, according to the Trinity doctrine, the three must share power equally—a 'horizontal power structure', which means no member of this 'club of three' is supposed to be superior to another. What one member knows, the other must also know. In other words, they must possess the same capabilities and powers, whether physical or intellectual, since they are 'supposedly' one entity.

From Jesus Christ's personal explanation in the Bible, the relationship between God the Father 'Elohim/Allah' and himself is hierarchal, a vertical relationship, contrary to the proposed horizontal power-sharing of the Trinity 'club of three'. According to this relationship, the Father 'Elohim/Allah' is superior to Jesus and every other entity in existence. This is referenced in multiple verses within the Bible, from the mouth of Jesus himself. As one scholar once remarked, the problem with the Trinitarian churches is: "You cannot take Jesus away from the mouth of Jesus". In other words, Jesus was not 'a deaf man' who required others to speak on his behalf. He publicly stated who he was and what his message was within the Bible, and it is readily available to any impartial eye to see it, as it is plainly recorded within the Bible.

Jesus calls God by the name of Eli/Eloe in the Bible, referenced in Mark 15:34 and Matthew 27:46. As we have established before, the name of God the Father to be and/or include (Eloh/Ella/Elohim) is widely documented throughout the Bible— over 2, 500 times. It has been referenced amongst other verses, starting with Genesis 1:1: "In the beginning, God [Elohim] created the heavens and the earth". After Jesus identifies who God is (Elohim/Ella/Allah), he teaches his true followers that God the Father ('Elohim/Allah') is greater than him, referenced in John 14:28: "...for [God] the Father is greater than I"—completely disagreeing with the equal power-sharing of the Trinity doctrinal teachings of the Church.

Furthermore, Jesus stated that the Father God ('Elohim/Allah') is not only hierarchically greater than him but also is superior to everything else in existence, referenced in John 10:29: "My Father, who has given them to Me, is greater than all; and no one is able to snatch them out of My Father's hand." This passage is a unique verse. Jesus is saying no one (i. e.,'the Trinitarian advocates') should be able to snatch my sincere followers and my message out of the hands of the Father God ('Elohim/Allah'). The question for you is: do you want to be 'snatched away' from the hand of Jesus and Elohim/Allah?

To those who entertain the doctrine of Trinity and thus give worship to Jesus, the Messiah says to them: Your worship of me is in vain; you are practicing manmade doctrine as the commandments of God, referenced in Matthew 15:9 "...in vain they do worship me, teaching for 'manmade' doctrines as the commandments of God". Not only is Jesus' Father his God ('Elohim/Allah'), but God to all of us 'people', as Jesus said to Mary Magdalene, perhaps his last words, teaching people that he is not God, referenced in John 20:17: "...Go to my brothers and tell them, 'I am ascending to my Father and your Father, to my God and your God." What Jesus is saying is beyond the requirement of any interpretation. He is teaching his last words that God is not an exclusive Father to him, but to all of 'people'. But also, God is his God as much as our God. There is not even a hint of a 'Trinity' power-sharing 'club of three' here. Otherwise, according to Jesus, since God the Father is Father to all of us in this claim under the Trinitarian proposal, we should also be included in the power-sharing deal of the Trinity. Since God is our Father too, from the mouth of Jesus, as such, the Trinitarian claim should upgrade us all to become 'Lords', which you already know would amount to a vain claim anyway.

Jesus was not only verbally teaching that he is not God against the 'Trinitarian doctrine', but was proving via demonstration that Elohim/Allah is the singular 'highest' goal of worship—'Allahu Akbar'—to Allah is the greatest target, as he prostrated to Elohim/Allah, referenced in Matthew 26:39: "Going a little farther, he fell with his face to the ground and prayed, My Father if it is possible, may this cup be taken from me. Yet not as I will, but as you will". Notice the end statement, not as 'my will' clearly proves the power of Jesus' will is unlike the will of Elohim/Allah, in that his will is ineffective compared to the will of God, Elohim/Allah.

Jesus' prostration offerings to Elohim/Allah were consistently daily occurrences, with great resemblance to the way Muslims pray in their daily prayers, referenced in Mark 1:35: "Very early in the morning, while it was still dark, Jesus got up, left the house and went off to a solitary place, where he prayed". This prayer was offered by Muhammad every morning at dawn before sunrise, which Muslims call the 'Fajr' prayer. Many Christ-loving Christians who wanted to give their prostration to Elohim/Allah asked their Church pastors if they should prostrate to Ela/Elohim, and the reply they received is as resentful as it is contradictory to Jesus' conduct.

One of these Christ-loving people who asked their pastor included, but was not limited to, a famous You Tuber by the name of Brandon Estes, also known as 'the Muslim cowboy', a Texan native from Texas, USA. He remarked in his accessible interview on You Tube that he shared his intention of 'prostration' with his pastor "Randy", and he got a chastising reply that he was held 'in contempt', which was out of nowhere. As he puts it, Mr Estes was told, "We do not put our heads on the ground like "Moslims." Ironically, Jesus put his head on the ground. Is it not like 'Moslims'? Even Pastor Randy should have admitted this since his rank of pastorship ought to have a profound knowledge of Jesus Christ.

Pastors like Randy either do not have proficient knowledge of Jesus Christ, whom the religion should be based on as its central figure, and consequently should not be teaching about him, or the other alternative is that they are misrepresenting the message and conduct of Jesus 'intentionally'. With that said, the good news is that Jesus was more capable and well-expressed in speech than any pastors or religious priests to misrepresent his message, for those with an impartial eye to see the truth, should they choose to do so!

Furthermore, this act of worship, 'bowing down' to Elohim/Allah, is beyond Jesus. All the Abrahamic prophets conducted themselves in this humbling manner before Elohim/Allah without any exceptions. Moses and Aaron made themselves humble before the Lord of the universe via prostrating to him, referenced in Numbers 20:6: "Moses and Aaron went from the assembly to the entrance to the tent of meeting and fell facedown, and the glory of the LORD appeared to them".

Furthermore, the patriarch Abraham prostrated to Elohim/Allah, referenced in Genesis 17:3-4: "Abram fell facedown, and God said to him, 'As for me, this is my covenant with you: You will be the father of many nations". This verse is not only a testament that Abraham prostrated to Elohim/Allah, but that no particular nation can exclusively claim his bloodline, as it says Abraham is father to many nations, not only one nation. Muhammad, descendant of Abraham through the bloodline of Ishmael, comments on the subject matter, hence remarking that the most prestigious act of worship to the Lord is prostration, as the servant is nearest to his master at such a junction. Therefore, ask anything you are hoping to receive at that position. He says in the authentic Hadith, Sahih Muslim 482: "The nearest a servant comes to his Lord is when he is prostrating himself, so make supplication (in this state)".

Coincidentally, all the people who have asked their Lord under prostration 'the big question', what is the purpose of life and what is the truth, all reported to have received clarity in their 'heart' and quest, which the immortal intelligence finds to be an indispensable tool to utilize for the journey. With that said, one needs to be cautious and generic about the name used to call God in this supplication; otherwise, you might call the wrong person. Therefore, the advice is: If you would like to take this journey, wash yourself at your purest moment, prostrate to God, and stay as generic as possible in calling God by any name and say: "Oh my Maker, the Maker of the universe, I sincerely seek an answer to my purpose of life, please respond to me."

Jesus, from his mouth and actions, yet defines to his followers that the Father Allah/Elohim has superior power over himself and others, as he says, I cast the devil with the finger of God Elohim/Allah, not Jesus' power and will, but the power and will of Elohim/Allah,   referenced in Luke 11:20: "But if I drive out demons by the finger of God, then the kingdom of God has come upon you". I pray that we receive the true kingdom of God. Jesus further exhibits the superiority of God Elohim/Allah over himself when he was asked, "When is the judgment day?" to which he replied, Only God knows; no angel or prophet knows this day, as referenced in Mark 13:32: "But about that day or hour no one knows, not even the angels in heaven, nor the son, but only the Father."

A similar question was asked to the Prophet Muhammad about when the hour or judgment day is by an ordinary person, and he replied to him, "What have you prepared for it?" The man said, "Nothing but my affection for Elohim/Allah and his prophets [including Jesus]." Then the Prophet Muhammad replied, "You will accompany [in that day] those whom you have affection for." This is referenced in the authentic Hadith, Bukhari 3688, and then the attendees of the event said, "This is the best thing that ever came to our ears—Love the merciful God, and you will be with Him".

Furthermore, another 'elephant' dispute between Jesus and the Church is the doctrine of 'Under-Grace', which is nothing short of a grave breach of Jesus' message. The doctrine of 'Under-Grace' is taught by the Church through the self-appointed Apostle Paul, who changed his name from Saul of Tarsus due to his questionable background as an open enemy of Jesus and his early followers by trading as 'a bounty hunter' of the blood and flesh of the early 'Unitarian' followers of Jesus to be tortured and killed. Yes, this statement may be unexpected to many of you, but it is the direct teaching of Jesus within the Bible. The plain truth is that not only is Jesus contrary and incompatible with the doctrine of the Trinity, but with the doctrine of 'Under-Grace' as well, regardless of what glamorous shroud it is covered with.

In order to attain the paradise of the Father Elohim/Allah, all one needs to do is believe that Jesus has died for his/her sins; in other words, someone else, 'a third party', paid the dues and did the work for you. This begs the immortally intellectual question: If God wanted to forgive mankind with his 'sovereign grace', why is He coerced to make someone else suffer in order to make the devil happy for the goal of ransoming humanity from his anger, as taught in the Ransom doctrine of Christianity?

It cannot be any more authentic to any impartial eye to see that Jesus himself answered the ultimate question, a direct question: How can one attain the paradise of God Elohim/Allah? In Matthew 19:16-17, "And, behold, one came and said unto him, Good Master, what good thing shall I do, that I may have eternal life?" Notice the asker addressed Jesus as God by calling him 'Good Master', and asked a direct question: What does one need to do in

201

order to attain eternal life, 'paradise'? This is a salvation question put to Jesus, and Jesus, as eloquent as he was, never needed someone else to answer for him, neither Paul nor the Pope, and most definitely not your local Pastor. Hence, Jesus, as you would expect, gave a relevant and direct answer to the question. As he rejected the first part of the question, in which he is upgraded to becoming God Almighty, 'Good Master,' he replied in the King James Version: So He said to him, "Why do you call Me good? No one is good but One, that is, God. But if you wish to enter into eternal life, keep the commandments". In the first part, he rejects yet again the idea that Jesus can be deemed to be God('Trinity') by saying, "Why do you call me Good Master?" In the second part, he is separating himself from God Almighty, the Elohim/Allah, by saying, "Only God is good 'not me'".

So far, up to that point, he only made corrections as to how to understand who Jesus is as opposed to who the Father Elohim/Allah is but has not yet answered what someone needs to do to succeed and win the prize of paradise, as this is the core question raised by the asker in this verse. Jesus, in reply to the question, said simply, "Keep the commandments." In other words, the requirement, the prerequisite to paradise, is keeping and upholding the commandments, quite the contrary of Paul's invented 'Under-Grace', which is taught and enforced in all Christian churches today without exceptions, regardless of denominations.

For Jesus, the salvation which he came to preach is clear: It is to uphold and maintain the applicability and enforceability of the law of Moses, known as the commandments of 'The Torah', starting with the first commandment, which essentially highlights the upholding of a Unitary All-Powerful God, unlike the Trinity God developed 300 years after Jesus. It says in Deuteronomy 5:7 and Exodus 20:3, which, in paraphrase, proclaims that God is a Unitarian God; hence, do not give any acts of worship to anyone else. Furthermore, the one God has no image; therefore, no pictures, images, statues, or Idol is applied into his warship, in doing so will give rise a heavy, protracted punishment from God Elohim/Allah to the practitioner: "You shall have no other gods before me. You shall not make for yourself an image in the form of anything in heaven above or on the earth beneath or in the waters below. You shall not bow down to them or worship them; for I, the LORD your God, am a jealous God..."

With the above clarity, the artificial orbit cannot deceive people with the invented 'Under-Grace' doctrine, except for those who choose to stay ignorant about Jesus. As attractive as 'Under-Grace' is, due to its 'impunity', which gives people a sense of carefree, sin-free lifestyle, it triggers the fundamental question: Who does the religion of Christianity belong to? Is it the Christ 'Jesus', whom the religion is named after, or is it Paul and others who are pursuing their own doctrine, dressed as a commandment from God? As Jesus so eloquently put it in Matthew 15:9, You teach people invented doctrines as a matter of commandment of God, thus they worship me in vain.

Further, Jesus warned against this doctrine of 'Under-Grace' beyond the requirement of 'interpretation' or the semantics of Biblical jargon in clear, unequivocal terms: Whoever is applying the 'doctrine of Under- Grace', adhering to it, or preaching it, will qualify to draw the full anger and wrath of his/her Lord, so much so that such a person will become the least person to God. In other words, being a participant in this doctrine is the worst thing you can do for yourself. This is referenced in Matthew 5:19: "Therefore, anyone who sets aside one of the least of these commands and teaches others accordingly will be called least in the kingdom of heaven, but whoever practices and teaches these commands will be called great in the kingdom of heaven".

Observe how Jesus is saying that not only are the appliers, i.e., those who accept this concept of 'Under-Grace' are, losers in the Kingdom of God, but also those who teach it will amount to being virtually 'nothing' in the sight of God. Hence, the opposite proves to be the highest and best in the presence of the Lord, which literally means the more you reject this doctrine of 'Under-Grace', the more you are successful and stand great in the 'kingdom of God', i.e., the closer you are to winning the paradise prize.

This is what the immortal intelligence would call the Shakespearean weighing-up moment, a dilemma of "To be or not to be". It is time to be sincere with yourself: Either you genuinely follow Jesus Christ, hence your identity of being Christian is authentic, or you follow the accountability-free, invented doctrine of 'Under-Grace' because it is a 'deal' too good to be true to pass, even if it may not originate from Jesus, nor was it endorsed by him. My dear friend, it has always been your free will choice—it is your call for your soul.

Another major overlooked mishap, perhaps the most grievous theological fallacy for the average layperson like myself and you, is how to understand the title 'Son of God' versus 'son of god' in the Biblical parlance 'lingo', and why Jesus is God on the one hand and the Son of God on the other, which translates to Jesus being His own Son as much as His own Father in the Trinity 'framework' among others. In the traditional Biblical jargon, which is supposed to be based on Middle Eastern Semitic cultures, the concept 'son of God' has been referred to and appeared as the description of good servants of God who simply are —people who comply with God, God-compliant people, God-obeying people, God-fearing people, God-conformists, those who submitted themselves to the Lord etcetera.

In its highest form, those who were conferred this title with the descriptive servitude to God were those who became 'at one' with their Lord. They are those whose obedience to God is not only the fulfillment of the commandment, but they agree that this is the right thing to do. Not only does the only real Sovereign King have the right to command, but His commandments are the appropriate decision, just, and prudent for the subject matter. Therefore, they totally accept the decree of God, hence being at one with Him. For instance: if God prohibits the usage of narcotics, it is not a mere exercise of God's supremacy, but it serves benefit to the people.

An example is conferred: The typical sons of God were a spectrum, beginning with the supreme pious community of people such as prophets, followed by their disciples, students, or companions, to even common folk who showed affection towards their Lord, who were also called sons of God, or in its plural form 'children of God'. Furthermore, the title was granted to angels as well as what 'appears' to us as inanimate objects, such as celestial bodies and stars. In the prophets' form, Adam is referred to as the son of God, referenced in Luke 3:38: "Kenan was the son of Enosh. Enoch was the son of Seth. Seth was the son of Adam. Adam was the son of God." Additionally, Jacob is also referred to as a son of God, referenced in Exodus 4:22: "Then you shall say to Pharaoh, 'Thus says the LORD: "Israel is My son, My firstborn". Solomon received a synonymous description of son of God, referenced in 1 Chronicles 28:9: "As for you, my son Solomon, know the God of your father, and serve Him with a loyal heart...."

In the form of disciples and other servants of God, referenced in John 1:12, Jesus informs the disciples that they qualify to be 'children of God' collectively due to their obedience and support of him. Furthermore, in Matthew 5:9, good people who promote and work towards peace are also called sons of God: "Blessed are the peacemakers, for they shall be called sons of God". In the form of angels being called the sons of God, this is also referenced in Job 1:6. God, the Elohim/Allah, summoned all the angels, and they assembled before His Majesty, including Iblis 'Lucifer', something we saw in the story of Genesis in the Koran: "Now there was a day when the sons of God came to present themselves before the Lord, and Satan also came among them". In the form of a celestial body, the immortal intelligence records the term 'son of God' going to the compliant creation of God, in this case 'the stars', referenced in Job 38:7: "When the morning stars sang together, and all the sons of God shouted for joy?" The immortal intelligence notes the interpretation of some Biblical commentary, which refers to the stars as angels, but dismisses this interpretation due to the fact that, if the stars can sing, it also means they can shout, which 'singing' entails as a form of shouting. Therefore, despite stars appearing to us as lifeless beings, they, in all likelihood, have their own life in a different dimension to us on the planet Earth. With that said, even if the metaphor is aimed at angels, the commentary does not change the facts presented above.

The aforesaid has always been the traditional understanding and applicability of the 'son of God' until a new concept, with the motive to enlarge the scope of the indivisible Unitary God, came into development. Subsequently, its parameters were extended while adding two completely new entities separate from Elohim/Allah, which will be identified as such because the Trinity to date is presented as three distinct persons, yet somehow these three are one. This comes down to diluting the first commandment – 'cheating it' would be the perfect fit for depiction, particularly its exclusivity, which is emphatically reserved for the 'Singularity' of God, which can neither be added nor subtracted.

In this new development, Jesus would be upgraded, without his consent, to be the 'Son of God' with a capital letter 'S', and the previous sons of God, enumerated in a large pool of people and angels, would be lowercase letter 's'. This would be achieved through the fact that Jesus was born miraculously without a father; hence, he could be presented to the masses, who were primarily 'Gentiles' (i.e., pagan Europeans and others at the time), as the only begotten 'Son of God', due to their lack of awareness and familiarity with respect to the Semitic culture and language in which the scriptures originated, with an absolute restricted Unitarian God without any compromise as the central message.

The immortal intelligence finds the above claim to be not only flawed and irrational but completely lacking clear and unambiguous backing from any authoritative primary source, such as Jesus of Nazareth himself. Is it not shocking that the very person whom the Trinitarians are promoting to be God or the only literal begotten Son of God, in verses like John 3:16, "For God so loved the world, that He gave His only begotten Son, that whosoever believes in Him should not perish, but have everlasting life", is debunking and disqualifying their claim, which makes this verse and its objective of

Jesus not only flawed but indefensibly redundant, from none other than Jesus of 'Nazareth' himself? He teaches his followers, God Ela/Allaha is my Father as well as your Father, My God as well as your God, in John 20:17. Jesus directly says to teach/tell my brothers/followers: "I am ascending to My Father and your Father, and My God and your God". There is no doubt that this was the last message Jesus intended to be transmitted from him, and it discredits the doctrine that Jesus is the literal begotten 'Son of God'.

The whole fiasco of creating the capital 'S' claim, which was to make Jesus of Nazareth the literal Son of God and the rest sons of God as lowercase 's' 'son of God', was not well thought-out. It was a short-sighted coup d'état to take over the message of Jesus on the part of the Trinitarians, as it backfires in a big way. This is because the Semitic languages, such as Aramaic, Arabic, and Hebrew, do not have upper/lowercase letters in their alphabets. Consequently, the concept is flawed and expanded the parameters of Jesus' teachings way after Jesus ascended to His God, who is 'our God' as much as His God, as Jesus so eloquently puts it. As Jesus once said to his arch-nemesis, the 'Pharisees and Sadducees', the high priests of the temple of Jerusalem (paraphrased), You strive yourselves in the way of the scriptures, claiming that you are upholding them. Those very scriptures you are claiming affirm that I am the Messiah, and yet you omit the admittance of that fact, as referenced in John 5:39 in the New King James: "You search the Scriptures, for in them you think you have eternal life; and these are they which testify of Me. "

Making Jesus divine was the aim for these Trinitarians, contrary to Jesus' teachings. That is why the first attempt at scrutinizing the authenticity of the Bible, the only verse which embodies the 'horizontal' equality between Elohim/Allah and Jesus, along with the Holy Ghost, infamously known as the 'only Trinity verse' in the Bible, 1 John 5:7, "For there are three that bear record in heaven, the Father, the Word, and the Holy Ghost: and these three are one," completely failed.

The Renaissance scholar Desiderius Erasmus from the Netherlands, who carried out this verification test of the Bible in the 16th century—the first of its kind at the time—quickly found that this verse was not part of the earliest manuscripts written in Koine Greek and, therefore, omitted it. However, the Trinitarian churches, backed by the powers of the Kingdom, coerced him to re-insert this verse after removing it in his third edition of the King James Bible in 1611. That is why this verse appears in some Bible books and not others. Its roots trace back to the fourth century Liber Apologeticus, which then found itself to appear in Latin and Vulgate manuscripts in the sixth century. This means that this theological 'Trinity' understanding of Jesus was conceptualized some 300 to 600 years after Jesus.

If, as seen above, the aforementioned evidence substantiates that Jesus is neither the Son of God nor God in a Trinity, a horizontal 'club of three' members, then what is he? This is the curiosity of all of us with immortal intelligence! Biblical verses verify that Jesus of

Nazareth was received by the people as a prophet of God, rather than as divine, hence they testify to that declaration, referenced in Matthew 21:11. The public attested that Jesus of Nazareth is a prophet of God: "The crowds answered, 'This is Jesus, the prophet from Nazareth in Galilee". You will see multiple verses proving the prophethood of Jesus within the current Bible, which the Trinitarian churches show a lethargic attitude towards, notwithstanding the fact.

Another piece of evidence for his prophethood is 'stronger yet' due to the fact that it comes from Jesus' own mouth, explaining who he is. As he said to his adversaries, the Pharisees and Sadducees, referenced in Matthew 13:57, "Prophets are denied their due honor only in their homelands: 'And they took offence at him. But Jesus said to them, "Prophets are not without honor except in their own country and in their own house". This verse, coming from the mouth of Jesus himself, directly is part of those passages which will gain precedence over descriptive verses where a third party is commenting on the nature of Jesus. As a result, Jesus attests that he is a Prophet of God.

The second thing it shows is that Jesus was meeting heavy antagonism and hostility from the scholars of the Temple of Jerusalem rather than from the general folk. This is because Jesus disapproved of the commercialization of business trade they commissioned inside the Temple—the mass subscription of society for profiteering. In the eyes of Jesus, these practitioners in the Temple were stealers from the lay folk of the general public, as he charged them with converting the House of the Father 'into a Den of Robbers'. He indicted them, saying that instead of teaching and facilitating people in the worship of the Singular God, they commercialized the house and turned worship into a wealth-generating market. This is referenced in multiple places within the Bible, including John 2:13-16: "..Jesus went up to Jerusalem. In the temple courts, he found people selling...sitting at tables exchanging money. So he made a whip out of cordsand drove all from the temple courts, both sheep and cattle; he scattered the coins of the money changers and overturned their tables. To those who sold doves, he said, 'Get these out of here! Stop turning my Father's house into a market!'"

Again, Jesus' own statement from his mouth is held in higher regard than the comments of others about him. Jesus clarifies that he is a prophet of God, referenced in Luke 13:33 NKJV "Nevertheless, I must journey today, tomorrow, and the day following; for it cannot be that a prophet should perish outside of Jerusalem." This establishes Jesus calling directly himself a prophet of God and therefore automatically sets aside any other claim, particularly if it is narrated or from a third party, regardless of who they are.

Jesus' prophethood is well established, even by the account of his archenemies, the organized clergy of Jerusalem's Temple. Including the chief priests, they testify to it, referenced in Matthew 21:45-46. The chief priests of the Temple, along with the upper management of the Judaism religion at the time, known as the Pharisees, who were complicit in harming Jesus, intended to carry out their goal but refrained due to fear of backlash from the public, as the 'people' embraced him as a prophet referenced in NKJV: "Now when the chief priests and Pharisees heard His parables, they perceived that He was speaking of them. But when they sought to lay hands on Him, they feared the multitudes, because they took Him for a prophet. "

Another piece of evidence to prove that Jesus was widely accepted as a prophet at the time is referenced in Luke 7:16. As the crowd of the public was struck with awe of Jesus, they glorified Elohim/Allah and proclaimed that God had raised a prophet from amongst us: "They were all filled with awe and praised God. 'A great prophet has appeared among us" they said. 'God has come to help his people". And how do we know it is Jesus they are talking about? Because the previous verse, 7:15, clearly mentions that it is indeed Jesus who performed the miracle of resurrecting a dead person in the community.

Thus, the immortal intelligence points to the irony that those who witnessed 'alive' Jesus performing the greatest miracle of resurrecting a dead person right before their eyes did not take him as a divine God, as they regarded him to be a prophet. How on earth, then, can those who have not seen a shred of miracles from him today, a man they have never met in the real world, let alone experienced his real-world miracles, blindly call him 'God'? This remains a mystery that does not make any sense; it is your call for your soul!

It should not be startling for Trinitarians to put forward the argument that Jesus is God and the Son of God, which means he is his own son and father at the same time. This is when the words lose their meaning to describe concepts such as the new phenomena which are keeping the traditional Western Civilizations under pressure to succumb. For instance, non-binary people who describe themselves as neither male gender nor female gender or meet Otherkins who describe themselves as non-human existences, although they are physically and biologically human. Or, as if the former two are not enough of mystification, they jointly formed a new phenomenon called 'intersectionality of the two', which is when the two phenomena blend.

This is explained as when the 'perceived' non-human identity of Otherkin becomes the 'perceived' non-gender identity of the Non-Binary 'in and of itself' and vice versa, which the immortal intelligence calls when the brain loses its meaning. Do not easily write them off, however, because the idea of the 'Trinity' as convoluted and as meaningless as it was, ultimately was successfully sold to the psyche of the population. Let me just repeat once more what was sold to the brain of generations of people: that God is One and Three; that is a Son who is his own father, who is also a disconnected Ghost (third person), separate from the other two, while remaining within one person somehow.

The same God has his biological mother, called Mary, who is the Son's mother but cannot be a mother to the Father, who is also the same as the Son because the 'Trinity' is One God. This idea was sold to many in the Western Christian world after centuries and thousands of years of relentless campaigning and lobbying, as convoluted and irrational the doctrine was. All the while, lacking any evidence from Jesus himself, yet somehow it met no 'substantial' mental objection or resistance.

The immortal intelligence gives an analytical review in relation to the Biblical text of the notion that Jesus is 'fatherless', which is a 'matter of fact', well-established reality. Then, how do we understand a fatherless 'human'? Can he also be divine? Is he not the only Son to the Father 'Elohim/Eloe' since he has no biological father? It may be a surprise to many and a disappointment to others to hear the fact that, in all the commotion you hear, all the Bible codes you are preached to, all the singing, dancing, and gatherings for the Lord and Saviour in churches, nothing—no verse, no passage, no saying, no claim—ties or links Jesus himself to divinity directly in a clear-cut, unambiguous claim from the mouth of Jesus where he says, "I am God" or where he says, "worship me." Nothing in the entire Bible can substantiate Jesus ever claimed divinity in a direct, plain language. Trinitarians use certain verses in the Bible to justify the divinity of Jesus, such as John 14:6-11;Jesus said to him, "I am the way, and the truth, and the life. No one comes to the Father except through me. If you had known me, you would have known my Father also. From now on, you do know him and have seen him. "

Jesus fulfils the traditional duties of Abrahamic prophets, as all of them said something of similar effect. The purpose and functionality of a prophet works similarly to an ambassador of a country to your city, who is to represent his Head of State. Let us take the example of the United Kingdom: If the High Commissioner (ambassador) of the UK in your country says to you, as a visitor visa seeker, that the only way to travel to the UK is that you must fulfil the requirements of my Kingdom through me to interview you after you fill your visa application forms, you would not for a second equate the High Commissioner with the King of the United Kingdom and Northern Ireland, King Charles III. And if anyone says to you that the High Commissioner (ambassador) is the same as the King, you would not readily accept it as a question of fact unless they are bribing you with living 'Under-Grace', which means a free pass to eternity. In that case, you might choose to give a blind eye to all the red flags you see. Well, guess what? What if the bribe you are relying on is bogus because you do not have a cashable 'cheque' from Jesus of Nazareth himself?

Trinitarians further attempt to use these verses for the goal of upgrading Jesus to be divine, referenced in John 1:1-2, which reads: "In the beginning was the Word, and the Word was with God, and the Word was God. He was in the beginning with God."

Again, Jesus, being the Word of God, is fulfilling his prophetic role on earth to represent God. How can God communicate with us except through 'words'? Just like the ambassador represents his master, the Head of State, then comes the phrase "the Word was with God", which perfectly fits the role of being with God, being close to God, and being a member of the party of God. Then comes the phrase "the Word was God, " a contradictory statement. How is it possible that the Word was with God, and then the Word was God? What does that mean? Was God with another God? Because if A (the Word) is with B (God), then A is different from B, which means A cannot be B. Therefore, "the Word was God" does not make sense.

The immortal intelligence derives from the scriptures that yes, Jesus in the Spirit form, also known as 'AKA' 'the Word of God', was with God. Then the Angel Gabriel, also known as 'AKA' 'the Holy Spirit', brought the blessed spirit of Jesus to Mary, where the Word of God was fused into a flesh, a human body as we know it. By the way, the process of human development in the womb is essentially the same for all humans, including Jesus. Angels fuse the unseen soul, also known as 'spirit, ' into the fetus. Make no mistake, not all spirits have the same standing. Jesus' spirit may be superior to many other souls, as referenced in the scriptures, but all spirits belong to God as His property. They are fused into the fetus, usually around the 15th-17th week mark.

However, in the 7th or 8th week of pregnancy, the fetus begins to show signs of movement, although partial. We may not equate the simple sign of movement with the spirit fusion because, at the onset, the sperm is moving to fertilize the egg, particularly in its subsequent zygote stage, in which it makes a significant journey. But we consider this existence a different dimensional existence at that stage, while the spirit—this mystery of us—was separate from the body. Therefore, the fusion is the process when the two are attached as one: 'soul and body'. Diffusion takes place at death when the two are separated, at which time the body remains while the spirit departs. This is known as the process of death, which means the duration of your test 'exam' is over.

In just a few verses after the one referenced above, interestingly and quite conversely, Jesus was asked a direct question as to what is his nature: "Are you the Son of God?" i.e., some kind of Divine or not? He responded, "I am the Son of Man", meaning mankind. Jesus was asked by a man named Nathanael in John 1:49, "Rabbi, you are the Son of God..." and Jesus corrected him in his response, "I am the son of man" in John 1:51: "Truly, truly, I say to you, you will see heaven opened, and the angels of God ascending and descending on the Son of man." This perfectly corroborates with the long-standing, well-established theological basis of all the Abrahamic books that God is not a man nor a child of man.

Henceforth, the two should never be blended; they are two separate beings in that one 'man'—is created, weak, prays to God, gets sick, waiting for his demise (death), and needs the sustenance of food and water to maintain his existence. While on the other hand 'God' is Omnipotent, possessing power over all things, Absolute Sovereign, which means He does not require rain or oxygen to maintain His existence. Also, God does not need to repent to anyone nor trespass against anyone; the dominion is His altogether.

That is why in Book Chapters 23:19 NKJV, it says, "God is not a man, that He should lie, Nor a son of man, that He should repent. . ." This coincides with the Koranic description of God in 112:3: "God is not a child to anyone nor a literal parent. " God teaches His sincere servants that God is neither human nor a child of humans, which Jesus taught the man named Nathanael. This is because, in all likelihood, Nathanael misunderstood by mistake about the nature of Jesus, possibly due to Jesus' 'fatherlessness, ' compounded by his consistent significant undertaking of miracles—just as many folks out there today also have similar misunderstanding. Therefore, if you are like Nathanael, you should take the teaching of Jesus of Nazareth seriously and call him what he called himself 'son of man'.

However, with that said, His 'fatherlessness' is still outstanding. Definitely, Jesus is different from normal human beings, despite his own claim from his own mouth that he is a prophet. This can be explained in terms of His miracles because other prophets performed miracles as well. Namely, Moses performed the miracle of parting the Red Sea, referenced in Exodus 14:21-22, or Joshua ordered the sun to stand still so that the darkness of the night did not come too early for them during a battle, referenced in Joshua 10:12-14. Isaiah also healed sick people, referenced in 2 Kings 20:1-11. Furthermore, Muhammad of Islam had undertaken some major miracles as well, namely splitting the moon during a debate challenge to prove his prophethood, referenced in Koran 54:1-2, or when he repaired a blind man whose eye had fallen off during a battle in front of people, referenced in Al-Tabaqaat-ul-Kubra li ibn Saad, Vol. 3, p. 239, or the book of proofs of prophethood by Dala'il al-Nubuwwah by Al-Bayhaqi.

While we are at it in this topic of prophetic miracles, it is interesting to relate the incident referenced in the authentic Hadith Musnad Ahmad 2166, where the Prophet Muhammad rejected performing a miracle in response to the demand of his people. They promised that, upon his doing the miracle, they would believe in him. The story is that the people of Mecca demanded that he convert the surrounding mountains in the city into gold so that they would believe in him. Although initially, he agreed and thought this great opportunity would offer salvation to his people, as he prayed to Allah to enable him to carry it out, Allah informed him that this was an additional, specific miracle for specific people. Should they not believe in him immediately after the miracle they demanded was performed, Allah's swift command would cease, revoking their license to live any longer—their 'lifespan' originally allocated to them as individuals.

Alternatively, Allah gave the option for them to continue living with their freewill 'right to reject' questioning and investigating the preset evidences already forwarded, which would permit them to complete their full lifespan granted to them at birth. Upon learning of the 'dire' consequences, out of love and care for his people, the Prophet refused to perform the miracle and backed out from his original agreement. Hence, imaginably, he was ridiculed for his backing out from it for some time.

This mockery of Muhammad's pragmatism, regarding his inability to carry out a miracle to prove his prophethood, found its way into Western discourse when the English philosopher Francis Bacon used the parable, "If the mountain will not come to Muhammad, then Muhammad must go to the mountain" in his Essays of Baldness in 1625, just before his license to life expired as he died within a year later in 1626.

Reverting to our discussion, despite other prophets who may have performed certain miracles, as well as Jesus, none of them is as fatherless as he is. This must count for something, and what could it be? We have ascertained thus far, at least according to Jesus himself, that he is not God the Father, nor is he the literal Son of God, as he called himself 'son of man', If you think these repetitive utterings of Jesus, amounting to over 70 times in the Bible, are in vain or a game, you are rather a fictionist fool who is making no sense to yourself. My dear friend, you have been caught in the groundless, accountability-free doctrine of "under grace'.

The fact is, this fatherless question the immortal intelligence finds not only interesting and rational but a legitimate question to inquire about. Henceforth, let us dive in and see what the Abrahamic scriptures say about the subject matter in conjunction with our natural intelligence. Despite the Bible foretelling that God is not a son or a father in human terms, which is —neither begotten nor begets, as referenced above in the book of Chapters— substantiating with the fact that Jesus repeatedly proclaimed to the masses, "I am son of man" over 70 times referred throughout the Bible, as we have seen above, even Jesus prophesying that people would give him vain worship via believing in invented, manmade verses in the Bible prescribed to them as a commandment from God in Matthew 15:9.

The full effect of the Trinity, with the miraculous birth of Jesus being 'in the thick of it', did not fully evolve until a couple of hundred years after Jesus ascended. This will be unveiled at a later stage with respect to the open secret evolution of the Trinity—who added what, when, and where?

Going forward, since the Koran is the only Abrahamic scripture that was revealed some 600 years after Jesus and 300 years after the Trinity doctrine was canonized, there will be direct and a head-on collision between the Koran and the doctrine of the Trinity, along with the interpretation of the fatherless human being called Jesus as the 'begotten Son'—unlike any other Abrahamic scriptures such as the Hebrew Bible, where Moses preceded the controversy.

However, we are omitting the New Testament in the contest, since we are hypothesizing that the doctrine was an aftermarket addition cultivated inside the New Testament.

Contrary to common misconception, the Koran is the only Abrahamic scripture where the direct pronoun of the Trinity is mentioned. Astonishingly, it is an amazing veracity that the entire Bible, including the New Testament, does not have the verbatim pronoun of the Trinity, whereas, on the other hand, the Koran does. It says, in simple, straightforward language in Koran 5:73 (paraphrased), "Do not agree with the claim that God (Elohim/Allah) is part of the 'Trinity'—you know this claim has no legs to stand on. There are no gods but One Singular God".

Our focus is not the Trinity itself at this juncture but the fatherless Jesus. The Koran makes it clear in unambiguous, explicit terms that the Trinity—three persons or Gods—are not compliant with 'Singularity, ' which means inseparable Unitarian God. It also sheds light on how to perceive the phenomenon of a fatherless human being, which even startled Mary herself after being notified of the prospect of having a child without a father. The famous chapter of Mary in the Koran chapter 19:17says, "We sent to her our Spirit". Notice the angel is given the title 'Spirit'. Henceforth, after getting over the initial shock of encountering the 'stranger', the angel told her, "You will have a pure baby boy" referenced in verse 19, to which she replied, How is it possible to have a boy when I do not have a husband? (verse 20). "She said, 'How can I have a son when no man has touched me, and I was never unchaste?'"

Observe, even to Mary herself, the prospect was unfathomable. It baffled her. So, if you are trying to scratch your head for the possibility of a 'fatherless' pure boy, remember, it is not only you. It is actually natural to feel that way because Mary herself felt the same. Please take note of how the angel responded to her question in verse 21: "He said, 'Thus said your Lord, it is easy for Me, and We will make him a sign for humanity, and a mercy from Us. It is a matter already decided."

"It is easy for Me,"Allah responded, reminding her that He is God of the entire universe, the Omnipotent, and has power over all things. In the early part, she was startled herself to grasp the possibility. However, after she gave birth, she experienced a different kind of anxiety: how would she break the news to people that she had a fatherless baby? "No one would believe me in the conservative community" she thought. In which God consoled her heart in verse 26: "So eat, and drink, and be consoled. And if you see any human, say, I have vowed a fast to the Most Gracious, so I will not speak to any human today."

In other words, God will answer that difficult question on your behalf to the people, and exactly her fears would prove to be true. As she came back to town with the 'strange' baby Jesus in her arms, people surrounded her and asked her how on earth this happened, referenced in verse 27: "Then she came to her people, carrying him.

They said, 'O Mary, you have done something terrible,'" making the presumption that she had an extramarital child—unlike her 'Mary,' the chaste, highly religious girl. If, at this point, she had responded to them, they would not have believed her, as they said, "What a dreadful thing you did, bringing shame to your parents, "referenced in verse 28: "O sister of Aaron, your father was not an evil man, and your mother was not unchaste".

At this intersection, she pointed at the few-days-old baby in her defense because her Lord told her, I will shield you from such a shame, as referenced in verse 29: "So she pointed to him. They said, 'How can we speak to an infant in your arms?'"

This is striking, to witness the insight of her trust in the Lord. Where most of us would falter, she maintained her faith and composure in her Lord despite receiving an early onslaught of appraisals towards her and her family. Nevertheless, she remained silent. It is here where Jesus performs his first miracle—by talking as a few days-old baby in defense of the honor of his mother and grandparents. Besides that, he reveals who he is: a 'prophet of God' a recurring utterance of Jesus by now. This is the reason why, in the Bible, it is referenced multiple times, including Matthew 21:11, that the multitude 'believed' in him and took him as a prophet because they already knew his first miraculous incident as baby Jesus gave a coherent, highly intellectual speech which quickly spread in the region like a sweeping wind referenced in verses 30-31 "He said, "I am the servant of God. He has given me the Scripture and made me a prophet. And has made me blessed wherever I may be; and has enjoined on me prayer and charity, so long as I live."

The **Koran** on the subject matter of the 'fatherless' Jesus reminds its readers to recall that the Lord created Adam, the first human being, without any parents. Hence, the creation of Jesus, the son of Mary, is similar to that of Adam in some respects, as referenced in **Koran** 3:59: "Indeed, the example of Jesus in the sight of Allah is like that of Adam. He created him from dust, then said to him, "Be!" And he was."

In conjunction with this, the **Koran** also reminds its readers that not only was Adam created without parents, but Eve (Hawwa), the first woman, was also created without parents, as referenced in **Koran** 4:1: "O humanity! Be mindful of your Lord, Who created you from a single soul, and from it, He created its mate, and through both, He spread countless men and women. And be mindful of Allah—in Whose Name you appeal to one another—and honor family ties. Surely Allah is ever Watchful over you."

It is this verse that scholars extrapolate the fact that 'the single soul' mentioned here refers to Adam, and from it refers to Hawwa (Eve). Hence, God the Singularity, the Creator of the Heavens and the Earth, wanted to demonstrate His creative powers through all four pragmatic possibilities of human creation. The first three are Adam, Eve, and Jesus, and the fourth is all of us. Since the fourth is so obvious, it is 'all of us,' as we come from two parents of the opposite genders: male and female. This is the first block of the existence of humanity, something the Non-Binary advocates cannot seem to grasp—the current 'effort' to dismantle this 'union' has a grave consequence for our human existential foundation, including themselves.

However, it is the relationship of the other three that is the target of our conversation. If the logic of Trinitarians has any validity, Adam would be the first God after the Father in the Trinity justification, because Adam would be supreme to Jesus due to the fact that he had no two parents—at least Jesus had a mother. This means Jesus, in all likelihood, went through the embryonic process that you and I go through, certainly including the pregnancy and birth pains that a mother experiences, but Adam did not. Furthermore, Adam began life as a 40-year-old adult. Hence, God created Adam in His image, that is, the image of adulthood, which means Adam was always looking like an adult. Whereas the rest of us, including Jesus (with the exception of Eve), went through the embryonic process and had different images at different stages—such as being an embryo at one stage, then a fetus at another stage, and so on—until we became toddlers and then reached our final adulthood anatomy at a later stage.

Also, as indicated by the above verse, Eve (Hawwa) was created from her mate's soul. Hence, the expression 'soul mate', the companionship of tranquility and partnership, which culminates in these two's union, has some scriptural roots after all. Eve, created from Adam, means that woman is created from man. As with everything, as soon as man is bestowed with strength and blessing, he may show arrogance. Perhaps due to man's arrogance, women suffer the implication of degradation in the Bible as second-class citizens cursed by God, as referenced in Genesis 3:16, among other verses.

The Almighty Allah wanted to debunk such a discourse beyond the point of doubt. As He created one of the most blessed people ever to set foot on earth—Jesus, the Son of Mary—from a woman, Mary, this simultaneously completes the fourth possibility of creation in order for the Almighty to demonstrate His Omnipotency. These four possibilities are: Adam created from nonentity, Eve (Hawwa)—her name in the **Koranic** tradition—created from male (Adam), the rest of us created from both male and female (our parents), and Jesus, the Son of Mary, created from a woman (Mary).

The point is clear: as much as God can create woman from man, God can also create man from woman. 'Do not get twisted'. That is why the Angel (also known as the 'Spirit') said to Mary, "This ruling of God has already been decreed and sealed", which means it must take place or come to pass.

Primary Evidence:
Science versus Bible

With all the above discussed, any book which claims to be from the Singularity, the unseen Creator of the heavens and the earth, the Omnipotent, the All-knowing, must back its claim with primary evidence before the immortal intelligence would even give the slightest consideration. If we give a quick recall, we said Jesus walked on water thousands of years ago would not suffice for today's scientifically and technologically minded community, who were not first-hand witnesses to such supernatural events. Similarly, Muhammad splitting the moon would not qualify as primary evidence for the 21st century's immortal intelligence. Hence, we may consider such claims as supporting or secondary evidence, but only after the primary evidence checks out. Therefore, let us analyze the primary evidence of the Bible as already forwarded.

The Age of the Universe

The Bible remarks on the age of the universe as being roughly 6,000 to 10,000 years, which is in direct conflict with scientific knowledge of the subject matter. As we extensively discussed above, the established age of the universe is 13.8 billion years old, with the planet Earth's age being 4.5 billion years old. Henceforth, if there is such a blatant miscalculation on the account of the Bible ascribes, it would amount to a disqualifying flag against the Bible being the pure word of God as claimed. This remark of a 'young earth' is rooted in the Bible in *Genesis 5-11*, which was calculated in the 16th century applying what became known as the 'Ussher Chronology'. This was based on Archbishop James Ussher's book *The Annals of the World*, published in 1658 and later verified by Dr. Floyd Nolan Jones's work in *The Chronology of the Old Testament*, published in 1993.

According to this chronology, Abraham was born 2,000 years after Adam, who was created on the sixth day. Consequently, Jesus was born 2,000 years after Abraham. This generates a timeline of 4,000 years distance between Adam and Jesus. Subsequently, from Jesus to the present day (2024), there is a time distance of 2,024 years. Adding this to the previous 4,000 years, we arrive at an estimation that Adam, the first human being, was created approximately 6,024 years ago as of 2024. Consequently, this would translate at least there cannot be any human beings who could have existed prior to that date of 6,024 years ago since Adam, the first, was created at that time.

However, the scientific community discovered human fossils and remains aged 300,000 years in a joint research effort by the Max Planck Institute for Evolutionary Anthropology (Leipzig, Germany) and Abdelouahed Ben-Ncer of the National Institute for Archaeology and Heritage (INSAP, Rabat, Morocco). These Homo sapien fossil bones, discovered in Jebel Irhoud, Morocco, in 2017, predate quite significantly, by 100,000 years, the previously oldest human fossils that were known to have existed up to that point of time which was discovered in OmoKibishin Ethiopia, in the year of 1967. These earlier fossils were dated to 195,000 years ago. It is clear beyond any reasonable doubt that 6,000 to 10,000 years is vastly different from the above dates—metophorically, a 'universe distance' away from 300,000 or even 195,000 years.

Therefore, the conclusion is obvious: the Bible scribes added their personal opinions into the book. This will become clearer as we scrutinize the Bible further. The shocking part is not the fact that people—'scribes'—added their own opinions into the Bible, but how readily this is coveredup, despite it being widely known to any Bible reader. Quite literally, without any exaggeration, it should be an 'open-secret' affair.

Humans speaking their minds in the Bible are plentiful, often without even identifying who they actually are in many instances. It is a recurring theme in the Bible, and their insertions, unfortunately, not only dent the reputation of the Bible as a devine book from the All-knowing God but, to many forward thinkers, completely disqualify it as inaccurate, unascertainable ancient texts. Their primitive scientific understandings, which are found in the Bible, is beyond debating point, whether incidental or intentional. Certain mistakes seem incidental, while others clearly benefit specific theological innovations. These insertions advance variant theologies, such as the Trinity, in order to expel Unitarianism within the Bible or depict Jesus as God the Almighty in order to overrun and disintegrate the prophethood of Jesus and by extension his authentic message as seen above within the Bible. Such insertions we will explore from here onwards in the sphere of science.

The Flat Earth

The Bible scribes suggested that the planet Earth is not only a flat Earth but a square one at that, hence, it has four corners that should be visible to the human eye. They wrote in *Revelation 7:1*: "I could see the four corners of the Earth, "which suggests that for the Earth to have four edges—'corners'—it must, therefore, be flat. This was the belief and scientific understanding of the day 2,000 years ago. This belief was further supported by early Egyptians, Greeks, and Babylonian philosophers, who believed the Earth was a flat instrument until Aristotle observed a spherical, round-shaped planet Earth.

One of the verses that fell victim to an uncertain scribe is none other than *Isaiah 11:12*, which claims in clear terms that the Earth has four corners, in line with the primitive cosmology of the flat Earth 'disk' taught in the world at the time, particularly in Babylonia. The verse suggests—although not definitively—that the Israelites would return to the land of Judah after exile from the four corners of the flat planet Earth. It reads: "He will set up a banner for the nations, and will assemble the outcasts of Israel, and gather together the dispersed of Judah from the four corners of the Earth." However, what is definitive is that the planet Earth is not flat and, as such, cannot have four corners unless the meaning of the four corners is to be taken as a complete figurative regard, which amounts farfetched.

What these kinds of verses show is that scribes and religious clerks have stooped to become intertwined with the normative political power chase of the day. When religious texts can be tilted for political power gain, that is why verses like this suffer multilayered contradictions. Not only is the statement not acceptable by the standards of scientific evidence—beyond any disputable opportunity that the planet Earth is flat—it also suffers from biblical-theological discrepancies.

It is a priori that the Israelites experienced multiple exiles—some ironically were holy and beneficial, while others were not holy or complementary. For historical reference, the first overlooked exodus occurred when the ten tribes, including the dominant Judah of today, decided to eliminate their younger brother Joseph this is referenced in Genesis 37:18-20; the Koran Yusuf 'Joseph'12:9 the bible version states NKJV "Now when they saw him afar off, even before he came near them, they conspired against him to kill him". Although, this only wound up becoming an extremely jealous rivalry between brothers by the intervention of God Elohim, hence prevented escalating to reach to a wicked murder. Nevertheless this unholy intent precipitated their first exile from the Holy Land to Egypt in the first place. However, the main exodus, which is well-documented in the Bible under the leadership of Moses from the tyrant Pharaoh, was a holy and complimentary event referenced in

Exodus 20:2. This was followed by the holy war waged by Joshua to enter the Holy Land, referenced in *Joshua 10:12-14.*

Derived from these verses and historical facts, both above incident's time frame the tribes of Israelites as people were not dispersed to the four corners of the world. When Jesus of Nazareth was not only rejected but also implicated to be harmed, despite his only intent being to bring good to the Temple and by extension the people of Israel, it reached a disappointing breaking point. Jesus ultimately turned away in what appears to be a 'permanent turn away', expressing despair for them from the compassionate Jesus. As he says in *Luke 13:34:* "Jerusalem, Jerusalem, you who kill the prophets and stone those sent to you, how often I have longed to gather your children together, as a hen gathers her chicks under her wings, and you were not willing!" This known compassionate man, whose care and love was so dismantled to a torturing level and reciprocated with hate and dismissal, would end up sealing the fate of the Temple. He stated to his disciples, as referenced in NKJV *Matthew 24:1-2:*

"Then Jesus went out and departed from the temple, and His disciples came up to show Him the buildings of the temple. And Jesus said to them, Do you not see all these things? Assuredly, I say to you, not one stone shall be left here upon another that shall not be thrown down."

Seemingly, according to Jesus, the Temple became far rebellious and outcast to God, so much so that it warranted no stone within it would remain standing in the hands of the Romans 70 AD 'After Christ'. Which meant its total destruction. This prophecy of Jesus against the Temple began to take hold and fulfilled. As a result, the demise of the Temple took place in 70 AD (After Christ). The destruction came through the wrath of God expressed in the tongue of Jesus of Nazareth at the hands of the Romans, not the Palestinians or Islam, nor the 'Arabs' if that is your preferred term.

The good news is that there is always hope of redemption for the sincere through repenting to God. As Jesus said in *Matthew 22:39*: "Treat your neighbor as you treat yourself". Which essentially teaches human equality and dignity should be provided to all, not only to your kind, no appertied or double standard to people who are different to you. Certainly, the pleasure of the Lord would not be earned through supporting, allowing, or participating in bloodshed and persecution, as enshrined in the commandments Jesus upheld "Thou shall not kill" in (*Exodus 20:13*) makes clear.

The immortal intelligence is pointing out these above documented peace-calling statements from Jesus, and those who amplify them, are the truest deserving of a Nobel Peace Prize from the Norwegian Nobel Committee if they are sincere about sustainable peace. This stands in stark contrast to physicists who 'on the other hand' dedicating their research to creating atomic weapons of mass destruction (WMD) for global domination—or, rather, global destructive capabilities—and those who incessantly fund such projects. It is no coincidence that the weapons sector is the most lucrative industry in the global market, breeding on the fear of war and conflict between neighbors, states and peoples.

The Chronology of Creation

Another scientific flaw is the chronology of creation stated in *Genesis* chapter 1. For instance, Bible scribes wrote in *Genesis 1:8* that the creation of day and night preceded the formation of the Sun and the Moon. "And God called the firmament Heaven. So the evening and the morning were the second day". This is not possible, as the light emanating from the Sun—or its absence—allows for the emergence of day and night as we know it, in relation to the rotation of Earth on its axis every 24 hours or so. This was first proposed in the heliocentric model by Nicolaus Copernicus in the 15th century and confirmed by Leon Foucault in 1851.

Remaining on the same topic, the Bible writers also stated in *Genesis 1:3,* "And God said, Let there be light: and there was light". That light was present on the planet Earth before the formation of the Sun and the Moon, as seen in 1:8 in the creation of 'morning and evening time frame'. This is viewed not only unscientific but irrational because the Sun and the Moon 'reflected' were the first source of light to the planet Earth, serving wide and beneficial purposes.

Furthermore, in the same above verse in Genesis 1:8, the scribes made the mistake of claiming that the creation of day and night came before the formation of the planet Earth; it is obvious without an Earth to rotate, there cannot be day or night as, affirmed by

Facoult in 1851, which further shows a flaw amongst the Bible scribes.

Again, the Bible writers colossally miscalculated the scientific process of 'cause and effect' when they claimed in *Genesis 1:11-12* that vegetation, trees, and plants were not only thriving but bearing fruit on planet Earth before the formation of the Sun (third day). It states in *Genesis 1:12* "The land produced vegetation: plants bearing seed according to their kinds and trees bearing fruit with seed in it according to their kinds. And God saw that it was good" and after this creation, four verses after Genesis 1:16 the forth day God created two lights the sun and the moon one for the day and for the night as if the moon is separate light "God made two great lights—the greater light to govern the day and the lesser light to govern the night..."

This clashes with modern science, which asserts that light is an essential component required for photosynthesis to take effect. This indispensable relationship between sunlight and plants was first discovered by the scientist Jan Ingenhousz in the 17th century and later elaborated on its functionality by Nicolas-Théodore de Saussure in the 18th century. This demonstrates how 'ordinary' people's opinions rather than God's words encroached into the Bible and were later presented as commandments of God, as Jesus described earlier in Matthew 15:9.

The Heavens are Held by Pillars

Perhaps the most obvious mishap in the Bible is the claim that the 'Bible writers' put forward in *Job 26:11* that there are pillars holding up the heavens (sky), similar to the ancient royal palaces or temples Bible writers were accustomed to viewing in those days, which could not be erected without foundations and pillars. The verse reads in *Job 26:11*(NKJV): "The pillars of heaven tremble, and are astonished at His rebuke."

Pillars Hold Planet Earth

Even more discrediting than the above is the claim that pillars are holding the planet Earth, as suggested by the scribes in *Job 9:6*. The concept of gravity and dark energy was an entirely oblivious notion to their primitive scientific minds, which proves extensive human infiltration into the words that would make up the Bible. *Job 9:6* reads: "He shakes the earth out of its place, and its pillars tremble."

Drinking Poison is not Hazardous to Followers

This, however, is the most outrageous human interference with the words of the Bible—perhaps with the intent to harm the unthinking fanatical worshipers as it prescribes to them drinking poison would not cause any injurious outcome of harm to true Christians. This is mentioned in *Mark 16:17-18* as it reads in (NKJV): "And these signs will follow those who believe: In My name, they will cast out demons; they will speak with new tongues;

they will take up serpents; and if they drink anything deadly, it will by no means hurt them; they will lay hands on the sick, and they will recover".

Some un-pragmatic zealous fanatics, whose reason is blinded by their own thoughtless enthusiasm, have acted on this verse, neglecting recklessly the defensive division of their intelligence. Therefore, ignoring as such what ought to be obvious to a sane human mind—that snake venom is harmful substance and highly likely could cause fatality—they allowed poison to infect their bodies in order to prove the lies within this verse are true, which they proved the 'opposite 'the truth of the lies in the verse were always the truth, and poison venom will only impact how toxin crumples the human organs and met their fatal ends in their own hands.

Such 'self-proclaimed' Bible scholars include Minister George Went Hensley of Dolly Pond Church of God, as well as others, including Pastor James Coots of Full Gospel Tabernacle in Jesus' Name Church in Middlesboro. Coots committed what was essentially suicide by acting out this verse as he was attempting to prove to the world its authenticity, he refused medical intervention, having faith in the verse, thus blindly hoping for the verse had given him immunity and protection from the effect of the venom. The only thing he proved was the counterfeit nature of the verse when he died. As a result, oh yes, they got him 'well', as they intent to get many more in different outfits and paths with the same end.

Infectious Diseases

Another bizarre human interpolation into the Bible with the same harmful consequences as that of the above, except this time, it has a greater scope to reach, is how to stop infectious diseases from spreading within and outside the household in the healthcare sector. It is unfathomable to even propose these kinds of nonsensical approaches to medicine in this day and age, in the book of *Leviticus 14:54-57* which puts forward the medical regime to treat infectious diseases.

"These are the regulations for any defiling skin disease, for a sore, for defiling moulds in fabric or in a house, and for a swelling, a rash, or a shiny spot, to determine when something is clean or unclean. These are the regulations for defiling skin diseases and defiling moulds..."

The Bible scribes thought that spraying blood in and around a house seven times would be a curative measure against infectious diseases and halt their spread. One could not get any wronger as 'fermented blood' that takes on and carrying microorganisms which carry and harness while harboring viruses and bacteria could exacerbate the spread of infectious diseases, causing far more harm and fatalities among the household and the general population at large, with the 'infected' blood acting as a vehicle to accelerate the process. This is referenced in *Leviticus 14:49-53*:

"To purify the house, he is to take two birds and some cedar wood, scarlet yarn, and hyssop. He shall kill one of the birds over fresh water in a clay pot. . . dip them into the blood of the dead bird and the fresh water, and sprinkle the house seven times..."

The immortal intelligence cannot help but point out the similarity this verse has with voodoo's 'black magic' practices of sacrificing animal blood to 'spirits' and the like.

Women Discrimination in Childbirth

The bizarre, primitive human minds interfering with what should have been the pure word of God persist. This time, we see discriminatory treatment against women for no medical reason other than to degrade them, can be in the following verse. In the 'postnatal' period, after mothers give birth, the Hebrew Bible claims baby daughters are less holy and clean than baby boys. As such, the postnatal period for mothers who give birth to daughters lasts twice as long than their counterparts, such as mothers who bear baby sons. A mother who delivers a female baby is deemed unclean for 80 days, compared to 40 days for a mother who delivers a male baby boy.

This is referenced in *Leviticus 12:1-5*:

"Speak to the children of Israel, saying: 'If a woman has conceived and borne a male child, then she shall be unclean seven days; as in the days of her customary impurity, she shall be unclean. And on the eighth day, the flesh of his foreskin shall be circumcised. She shall then continue in the blood of her purification for thirty-three days. She shall not touch any hallowed thing nor come into the sanctuary until the days of her purification are fulfilled. But if she bears a female child, then she shall be unclean two weeks, as in her customary impurity, and she shall continue in the blood of her purification sixty-six days..."

Women's Discrimination in Infidelity

Another deplorable statement in the Bible is how the High Priests—bear in mind, these are the same Priests of the Temple of Jerusalem whom Jesus called 'the sons of the devil', who carry out the works of their father, 'Satan', in *John 8:44*—claimed to possess some 'wealth-generating' supernatural powers of a bizarre nature. If there is an unfaithful wife within a marriage, the suspecting husband is instructed to pay 'them'. Due to their supposed supernatural abilities, they could detect an adulterous woman by offering her 'contaminated' water accompanied by the priest's curse. Should the poor woman fall ill—which is highly likely due to the contaminated water—she would then be found guilty. This is referenced in *Numbers 5:11-31*:

"...'If a man's wife goes astray and is unfaithful...then he is to take his wife to the priest. He must also take an offering...Then he shall take some holy water in a clay jar and put some dust from the tabernacle floor into the water...May this water that brings a curse enter your body so that your abdomen swells or your womb miscarries...'"

The Bible, as seen above—whether the New Testament or the Hebrew Bible—contains countless scientific, mathematical, and genealogical discrepancies. These clearly confirm to any impartial eye that not only did human writers interject their opinions and biases, but they were also incoherent among themselves as they often contradicting each other. These contradictions surpass at least hundreds or even thousands, depending on how one counts.

One primary example is the genealogy of Jesus himself, which is given in two opposing line of names 'accounts'. For instance, in *Matthew 1:6-7*, Jesus is listed as the descendant of Solomon, the son of David, whereas in *Luke 3:31*, Jesus is listed as the descendant of Nathan, the brother of Solomon, who is also a son of David. The infamous Professor Herman's contemptuous statement is applicable here: "Which is it?" Make up your minds, Bible writers!

Jesus versus Trinity

Unless you are among those who are callous 'couldn't careless' about the doctrine of the Trinity, so long as you are receiving the attractive deal of being 'Under-Grace'—even if it is accomplished at the expense of Jesus and 'the truth' as he stands for upholding commandments—you should already be puzzled, just like millions of others. As referenced earlier, the Trinity does not work, even from an abstract logical point of view. For instance, how can a Son be his own Father while also being a Ghost on the other hand? How can three separate persons somehow add up to be one person? Furthermore, the Son also has a mother, which triggers the question what would her relationship with the Father 'look like?' Who is her son, whom she is 'supposedly' parenting, which bursts open floodgates of unworkable differences which are incapable to make sense to the average intellectual human 'faith or no faith'.

In this section, the immortal intelligence will examine the known history of such a doctrine's evolution. Yes, it has evolved—not over years, but centuries—before becoming the convoluted final product it is today. We will also shed light on where the idea of the Triune God, or 'Trinity', was practiced prior to its canonization and enforcement across the territories of the Roman Empire.

The concept of a Triune God, or 'Trinity'—a God with three heads—was neither entirely novel nor unique to Christianity, contrary to common misconceptions. Its roots trace back to pagan practices in Eurasia through ancient religions such as Mithraism and

Saturnalian beliefs. These were prevalent not only as religious practices but also as ritualistic cultures with a stronghold in the Greco-Roman world.

In Mithraic theology, the Trinity appears in this 'mystery cult' through its cosmogonic nature as the heaven, the earth, and the oceans, which were known as the supreme triad of the Mithraic Pantheon. Documented evidence shows this religion was heavily practiced within the Roman Empire before and after Jesus, lasting for three to four centuries AD. Emperor Aurelian even made it Rome's official religion, erecting a prominent temple in Rome named after *Sol Invictus*.

The similarities are so apparent, so striking : that Mithras was the Sun-God, akin to the Trinitarian Jesus placed in this doctrine and later enforced in Christian theology. The day of worship was Sunday—the day dedicated to worshipping the Sun—hence, 'Sunday', it was also known as 'Sol Invictus', which means 'the unconquered Sun', a Roman pre-Christian festival. In this pagan religion, Mithras's birthday was celebrated on December 25th, which later was transitioned into the birthday of Jesus Christ in what became known as the mass and celebration held for Jesus in pagan birthday of Mirtha, henceforth Mirthas by substituting Jesus transformed *Sol Invictus* festival into the Christmas festive period as we know today. In all intents and purposes, Mirthas wearing the cloth of Jesus was sold to the Christ lovers worldwide, if the Devil wears Prada in

Lauren Weisberger's story, then Mirthas wears Jesus in Christmas in his birthday!

Historical academics in Christian theology suggest St. Paul and other New Testament writers incorporated some of the ideas of this pagan cult into their writings, such as Dr Martin Luther King Junior in his writing page 163, *"Light on the Old Testament from the Ancient East"* in 1948........."

"...after being in contact with these surrounding cultures and hearing certain doctrines expressed...When they sat down to write, they were expressing consciously that which had dwelled in their subconscious minds."

The immortal intelligence might add to that fact 'figuratively' even Ray Charles could not miss seeing these similarities, perhaps inspiring a recalibration of his renowned song: *"Hit the Road, Thrice, Don't You Come Back No More"*. Dr King's statement suggests that he had intended to reveal the fact that he was aware the 'purity' in the message of Jesus in modern Christian theology was contaminated by other cult practices in the vicinity, hence rendered subtle but precise hints under the circumstances in his writing.

The evergreen tree and its decorations during the pagan festive period of Christmas season also link back to Sun worship. During the dark, short days of the winter season in December in northern Europe, in particular with rampant infectious diseases such as winter flu, among others, causing higher rates of death would elicit people to be gripped with fear of sickness and untimely death. This would coerce the common folk to connect with a higher power.

In a pagan Roman European world where the Sun is worshipped and therefore its power is diminished in this time, the lay common pagan folk believed when God the Sun is feeble in December, its powers live through the Evergreen tree due it is ability to be green all seasons hence turn their warship in this period to this tree as the tree believed to symbolize immortality, was linked to the 'Holly' or 'Hollin' tradition, signifying that the Sun-God's power had not died completely and continues to live through the tree. To honor this, pagan worshippers hung parts of it on their doors for protection and brought the main tree inside their homes, decorating it as a form of sacrifice and price to pay for protection.

This ritual persists today among many who claim to follow and love Jesus in the name of Christ and Christianity despite its well-known pagan origins. If you are vulnerable person, getting your winter 'flu jab' offers better protection than 'a mere' tree with glitter balls hanging on its branches. If you are scientific minded and not performing the ritual for protection, then what are you doing it for? Tradition? Reflect on whose tradition it is, knowing that Mithras is wearing your Christmas—a self-reflective question for your soul, It is your call for your soul!

Saturnalia also impacted Christianity. It was a Roman pagan religious worship and festive offered in the winter solstice to the god Saturn, who was deemed the god of agriculture and harvest. As winter reaches its peak and then would turn the tide for spring to come, which meant better weather, this festival, coupled with its complimentary worship of Bacchus (the god of wine and celebratory mood). Saturnalia was the precursor of the 'Sol Invictus' of Mithras, which would commence from December 17th to December 23rd, in which many of today's rituals in the west are based, such as the festive mood and drunkenness that follows, many parties and gatherings, exchanging of gifts and the like which takes place just before Christmas.

Another deceptive personality during the festive season is Santa Claus just as Mirthas is covered with the name of Jesus, the Santa Claus, which by name represents St Nicholas, is substantially representing the Saturn god whose image was the elderly man with a white beard and red cloth. Christmas, due to its background in pagan practices of the past, many churches tried to outlaw its rituals but failed, including Puritans in England in the 17th century.

Besides the Puritans, prominent Christian scholars, including the renowned Hans Küng, have called for proper reformation of the Catholic Church to revert to the original Jesus of Nazareth as 'the right hand of God', enshrined in *Acts 7:56* and *Hebrews 10:12*. This portrays Jesus as a blessed man elevated by God through prophethood and messengership in the Gospels, rather than a convoluted Trinitarian Jesus that can be easily refuted by Jesus' own statements recorded in the Bible.

The Trinity is not only incoherent and unworkable as a formula but also invalid as theology due to the fact that it was not an original revelation by Jesus of Nazareth or any prophetic figures such as Moses or Abraham, so much so that many Christian sources admit this fact. For instance, the Harper Collins Bible Dictionary describes the Trinity as something not revealed in the original New or Old Testament of the Bible. This is nothing more than a type of admission that it was later developed. When the new Catholic Encyclopedia writes, "the formula itself does not reflect the immediate conscience of the period of origin", or during the era of 'early Christians' when Jesus 'himself' was calling on his message, the formula was nonexistent from the awareness of the masses 'multitude' as referenced earlier. "In the early Christian period, there was not yet a fully developed formula for the Trinity". Consequently, it is a clear acknowledgement that the doctrine was developed at a later stage.

"The formula itself does not reflect the immediate conscience of the period of origin, "

and confirms that during the era of 'early Christians', when Jesus himself was propagating his message, the Trinity formula was completely absent from the consciousness of the multitude:

"In the early Christian period, there was not yet a fully developed formula for the Trinity. "

To an impartial eye, the above means that the Trinity was not revealed but rather formulated and developed by others—not Jesus, not Moses, and not Abraham. Furthermore, main Church sources confess the fact that it took more than two centuries AD (after Jesus) at least for the doctrine to become what it is today. This will be discussed further below.

The immortal intelligence would like to highlight that what was developed was not entirely new entities within the Trinity. Oh no, my dear reader. The Father (*Elohim/Allah*), referenced in *Genesis 1:1*, already existed. So did Jesus the Prophet, referenced in *Matthew 13:57*, and the Holy Spirit, often described as an angel. Hence, the description of the term 'ghost'—because angels are unseen creatures to our human eye, as referenced in *Hebrews 1:14*, which calls them 'ministering spirits' with the purpose of aiding and protecting the good servants of the God Elohim. Which is one of the functions of angels.

Furthermore, God's prophets—including those with supernatural characteristics such as Adam and Jesus, both having no human fathers, along with the Holy Ghost being part of the angels who are unseen creatures —existed separately from each other throughout the scriptures. Hence, there was nothing new about these figures. However, what was developed was their aggregation into one deity, putting them together in one basket, 'one God' the Almighty. Thus, the doctrine was more of an innovation than an entirely new invention altogether for a good reason.

This innovation was precisely risen this way because it would have been too risky to introduce an entirely new deity which carries no trace in the scriptures, as it would have been quickly detected and thus be at risk of failing. For this reason, the Trinity was adopted to dilute and compromise the strict oneness of the Creator, as described in the first commandment of the Old Testament as well as the Koran. By doing so, it caused people to breach their first commandment 'contract' which means if your work is directed to other than your authentic Master (*Elohim/Allah*), then my dear reader, it would be unfair for the Master to acknowledge or recompense that work which went to others, particularly, as the Almighty does not share His spot with anyone else. This is readily apparent in the first commandment. 'checkmate'.

As a result, any good works performed by Trinitarians—whether internal, such as loving the merciful God, or external, such as giving charity not in His exclusive name—are rendered null, as the worship is not pure to Him.

The evolution of diluting the purity of the absolute singularity of *Elohim/Allah* began with the elevation of Jesus to deity status without his consensual teachings, as proven previously. Therefore, this marks the first stage of evolution, which was a 'duality' of God rather than an outright Trinity. The power of God moved from the Singularity of the Father, as Paul managed to make it share with the son Jesus.

Thus, this strand, which is aimed at extending the one God into two beings, can be seen in early works of St Paul's letters, such as *1 Corinthians 8:6*:

"Yet for us, there is but one God, the Father, from whom all things came and for whom we live; and there is but one Lord, Jesus Christ, through whom all things came and through whom we live. "

The absence of the third member of the Trinity in the essence of God is evident in this passage. Paul's primary focus was to cause a merger between the Father (*Elohim/Allah*) and Jesus the Prophet, dissolving the demarcation line between the two until full integration was reached. Consequently, Paul, at this stage, would mention the Holy Spirit not as a separate deity within the Trinity beyond the Father and Jesus therefore the Spirit seems an angelic being that would support good servants of God more in line with the traditional understanding of angels being 'ministering spirits' in *Hebrews 1:14*, who strengthen the believers of God (*Elohim/Allah*).

This is evident in Paul's mention of the Holy Spirit in *Romans 8:9*.

"You, however, are not in the realm of the flesh but are in the realm of the Spirit, if indeed the Spirit of God lives in you. And if anyone does not have the Spirit of Christ, they do not belong to Christ. "

This is the Holy Spirit that Paul is teaching at this stage, and it cannot be the third entity which is included with the Trinity as such because the spirit mentioned herewith is dual as well 'Spirit of God' and 'Spirit of Jesus' there is no room for third entity or option available within the scope of the above statement by Paul regarding the nature of the Holy Spirit.

From what can be observed in Paul's early work, he laid the ground work to dismantle the teachings of Jesus that he is distinct and separate from *Elohim* and their relationship is that of hierarchical, not horizontal, as Paul extended God from singularity to duality which amounts to multiplicity according to the first commandment because to *Elohim/Allah*, it makes no difference if you were to share His position with a single extra deity or a thousand— that is the crossover between monotheism and polytheism 'checkmate'- it all amounts to a grave violation of the first commandment in the same amount hence you would be found liable.

All Christian historians agree that the complex Trinity as a term was coined by Quintus Septimius Florens Tertullianus, better known as 'Tertullian', in 220 AD—over two centuries after Jesus. With a toxic background relative to Jesus, Tertullian not only grew up as a pagan (quite the antithesis to the unitary God) but was also a lawyer who specialized in 'propaganda and narrative creation' in what is known as 'rhetoric lawyer' perhaps the so-called unethical lawyer or immoral lawyer 'the hired gun' as in our vernacular of today. It was him, Tertullian, who conjoined 'glued' these three entities into oxymoronic unity. On the one hand, they are claimed to be one entity in the ordinary meaning of the word because it has to be in line with the Singularity of *Elohim/Allah* in the first commandment. However, Tertullian ensured that, on the other hand, these three parties are to be defined as three distinct persons:

the 'Father, the Son, and the Holy Spirit' unbondable wires. In what he called *"una substantia, tres personae"* - one substance but three persons.

The Immortal Intelligence poses this question to the audience: Did Jesus of Nazareth, the eloquent Messiah the Word of God, needed a lawyer to advocate for his rights to be 'a member of the Trinity club' some two centuries later in a different language and place? Because Turtillian was a Tunisian 'Charthage' lawyer stationed in North Africa. Is it possible that Jesus had a communicational defect to declare what his position is? The answer is obviously a resounding no. Hence, Jesus explained the matter and nature of God head-on clearly and unambiguously, surpassing no less than three times in the New Testament. None of these three times does he ever mention himself as Jesus or the Holy Ghost as sharing any power with the singular God, *Elohim/Allah*.

Interestingly, these three instances occurred when Jesus was directly asked about the nature of God. For example, in *Mark 12:28-29*:

"...he asked him, 'Of all the commandments, which is the most important?' 'The most important one,' answered Jesus, 'is this: Hear, O Israel: The Lord our God, the Lord is one 'Echad'.

The reason Jesus did not mention any Trinity is simple: there was no Trinity at that time or any time before it. This is the reason why the Christian historians reluctantly declare that the Trinity was not in the consciousness of the first generation of Christianity, which means it was later added. These questions were put to Jesus who answered them in a similar effect and therefore precluded himself from the Singularity God as recorded in *Matthew 22:37* and *Luke 10:27*.

Trinitarians attempt to legitimize their misleading doctrine by exploiting linguistic semantics, even though it has been widely proven that *1 John 5:7* ("For there are three that bear record in heaven, the Father, the Word, and the Holy Ghost, and these three are one") was an unauthentic addition and subsequently expunged. The only other verse that attempts to place these three on equal footing is *Matthew 28:19*, which reads:

"Go therefore and make disciples of all the nations, baptizing them in the name of the Father and of the Son and of the Holy Spirit. "

However, the same event seems to be recorded with contradictory accounts in *Mark 16:16*, where there is no mention of the Trinity:

"He who believes and is baptized will be saved, but he who does not believe will be condemned. "

When these above two do not work as hoped for, the Trinitarians would attempt to legitimize their misleading cause through John 14:19, which reads,"......He who has seen Me has seen the Father..."

However, the issue with this ambiguous verse is it goes against Jesus's direct saying in *John 5:37*, where he clarifies himself as an emissary who is a messenger of God 'who is sent me' and then makes it clear in a explicit terms, stating that no one has ever seen God:

"And the Father Himself, who sent Me, has testified of Me. You have neither heard His voice at any time nor seen His form". what these verses show is that the Bible is full of contradictions, it is unfathomable to regard that Jesus was a confused man to cause these above disagreements, but it is evident that many scribes had intended to change the true message of Jesus as he 'himself' attests to this fact that they did not believe in him nor embraced him referenced *John 5:38* in which Jesus says you do not have the pure words of God in your 'hearts' because you rejected his messenger who he had sent to you which is himself NKJV **"But you do not have His word abiding in you, because whom He sent, Him you do not believe".**

Oh readers, it is critical not to be confused by or caught up with the fact that these three entities—the Father (*Elohim/Allah*), Jesus the Prophet, and the Holy Spirit the angelic being—exist separately and independently throughout all the Abrahamic scriptures of the Bible and the Koran, but the point of contention lies in proving beyond reasonable doubt their coequality, co-substantiality, and synonymity, as the Trinity asserts with dogmatic faith justifications.

Thus, every verse that carries their names, mentioning these three within the same passage, may not give rise to establish their equality as such, the immortal intelligence affirms their existence separately and independently from one another. For instance, every professional sports team has an owner, a coach, and a player. This does not mean they are co-equal in their capacity and roles within the team. The owner reigns supreme over his employees all, including, managers 'coaches', and players, which is the reason why Jesus is informing the people that **"the Father is greater than I"** as referenced above. The father who is the owner in this case.

Why a lawyer like Tertullian was needed to create rhetoric which extends the scope of the unitary God enshrined in the first commandment and therefore emerged the doctrine of the Trinity, which inserts the Holy Spirit within it as a new member building on the work of Soul/Paul? Further, why did the four Gospel writers such as —Mark, Luke, John, and Matthew—conceal their last names and identities as a whole?

Should it not be shocking to you that the New Testament with the absence of St. Paul, are books with unknown authors, and unfortunately, the only one who is known author is Paul, whose original name was Saul of Tarsus. Saul was well-known and notorious for his hostility towards Christ and his followers as he was known as Soul, earning the title 'Saul the Persecutor of Christians'. This justifies why these people have the clear motive to impede the message of Jesus, if the train cannot be entirely stopped on the whole, they sought to derail it so that it would not reach its intended destination of pure worship offered to the singular God as it were. Not only did they show to have the motive and intent with changing names and further concealing their identities but they also executed their plans, as seen in the examples above.

Tertullian, 'the narrative creator', and other Trinitarians had a problem on their hands, which was: a public statement made by Jesus of Nazareth that could not be confronted head-on by repealing or retracting because it had already spread to the disciples and with its knock-on effect through them spread to the larger society—the 'multitude'. As a result, Once undesirable information falls into the hands of the public domain, it means it cannot be blocked from spreading rapidly any longer; thus, it becomes well immersed into the society, the only way to halt it is to hijack its meaning and then force the desired interpretation onto the people by penalizing the original meaning while rewarding the new interpretation, with the hope that, over time and through successive generations, the desired interpretation will overtake and supersede the original meaning. As a result, the new generation of 'Gentiles' who are disadvantaged due to their lack of familiarity with the original Jesus of Nazareth— whether it is his spoken language of Aramaic or his Semitic Middle Eastern culture.

This is precisely what happened over the 400 to 500 years following Jesus. Unlike the European 'Gentiles', the underlying reason Jesus could not be presented or 'sold' as the Almighty God to any of the Semitic cultures (mainly Muslims and Israelites 'Jews') to date is due to their profound familiarity with the singularity of *Elohim/Allah* which warrants no possibility of modification and/or alteration conducted on its essence of being unique and 'inimitable' singularity.

Therefore, what was the 'public' statement made by Jesus to his disciples that precipitated the creation of a new god—the Holy Ghost—to be compounded onto the singular God? It was Jesus' declaration to his disciples, and thereby to the public, about the coming of a new person who would carry on the works of God. The identity of this 'new person' became the subject of a 'battle of narratives', as the interpretation would shape the trajectory of belief. The stakes were high, as the singular God partially expanded by Paul—from a unitary being to a duality—now faced further redefinition.

There are no less than six descriptive adjectives given to this new person Jesus said would come after him the Councilor, the Helper, the Advocate, the Comforter, the Spirit of truth, and finally, the Holy Ghost, the one if you are a contemporary Christian, you would often hear more than any other name because it plays nicely in the hands of the Trinitarian view of God and strengthens it for which purpose it was created in the first place.

In *John 16:7* (*NKJV*), Jesus states:"But I tell you the truth: it is to your advantage that I am leaving; for if I do not leave, the Helper will not come to you; but if I go, I will send Him to you".

If you think Jesus could send the Helper directly without the permission of the Father (God), this would be a misunderstanding. Jesus clarifies in *John 14:16-17*:

"I will 'pray'—ask the Father to send you another Helper. . . "

Thus, there is no debate whatsoever from both sides —whether among Unitarians or Trinitarians—that someone sent by God would come after Jesus. This is evident in all the verses where Jesus declares the coming of a new person 'helper.' Such verses include *John 14:16-17, John 14:26, John 15:26*, and *John 16:7*. However, the contest has been identifying who is this person Jesus announced, especially in retrospect since over two thousand years elapsed when this statement was made by Jesus.

The Trinitarians sell the idea this person is a novel custom 'a new practice' by God Elohim, a Holy Spirit 'Ghost' to visit every person who embraces the Trinity doctrine, however, the Unitarians assert this person must be a human prophet in line with the traditional methodology of all Abrahamic scriptures never Eloh/*Allah* ever sent a 'Ghost' to warn people; Noah was a human prophet, Abraham was a human prophet, Moses was a human prophet, so as David and Solomon, and Jesus was a human prophet from his own mouth declaring I am a son of man, this is further substantiated by the fact God does not change his ways and policies referenced in *Malachi 3:6*:

"For I, the Lord, do not change..."

Consequently, why would God 'suddenly' change in the mouth of an open enemy to early Christians, the known persecutor of Unitarian Christians 'Saul of Tarsus' AKA 'St Paul' with anonymous bible writers who are likely his accomplices along with 'communication' lawyer Tertullian who is probably 'hired for money' as he himself never to have met Jesus in his lifetime- an established fact. Furthermore, why the God who was only One God with no 'member of the three club.' Trinity would end up sharing his position with multiple distinct persons as the 'artificial orbit' makes it as clear as the sun shining these three in Trinity are three different and distinct persons. Nothing will stop you in your free will to believe whatever you want, but you cannot justify to your sane brain that three distinct beings are one person. It is your call for your soul!

Returning to the contested pronoun of the 'Helper': Did Jesus mean a novel methodology of 'spirit/ghost', or the traditional understanding of perhaps another 'human prophet', therefore we ought to examine the meaning of the word of 'Spirit of truth' which is the precise pronoun contested here which describes the post-Jesus bearer of the message of the One God *Eloh/Allah*, as other names such as the Councilor, the Helper and the like do support the notion of human prophet as supposed to the ghost 'Spirit' the Trinitarians are perusing.

It is clear historically, the word spirit is used in the Bible as the 'unseen creatures' like angels referenced in *Hebrews 1:14;Genesis*

19:1-2 and by extension, Jinns and demons and evil spirits referenced in *Samuel 16:14, Mark 5:1-13, Matthew 8:16* and so on as well as human flesh such as prophets which is our focus at hand, particularly relating to our topic of who is taking over from Jesus after his departure as the message bearer of God as Jesus forewarns people that there will be fake prophets/apostles 'Antichrist' of bad 'spirit' without a doubt, which he means human beings here referenced in *1 John 4:1-3*:

"Beloved, do not believe every spirit, but test the spirits, whether they are of God; because many false prophets have gone out into the world. By this, you know the Spirit of God: Every spirit that confesses that Jesus Christ has come in the flesh is of God, and every spirit that does not confess that Jesus Christ has come in the flesh is not of God. And this is the spirit of the Antichrist, which you have heard was coming and is now already in the world".

As seen above, when Jesus says there is Spirit of Truth that will take over from me, he means a good prophet rather than what he called 'false prophets'. The fact that he says 'test them' means do not take them at face-value every apostle who self appoints himself such as 'Paul' and 'Tertullian'.

In light of the above context, now Jesus says there will be a 'Spirit of Truth', a message bearer of God *Eloh/Allah* that will come after me with these four signs now this would become the ultimate test for this Spirit of Truth as Jesus required his followers to carry out:

1. He will make things that you cannot bear now clear to you.

2. He will not speak of himself; therefore, he will speak what he hears—'Revelation' from the Father (*Eloh/Allah*).

3. He will speak of me in high regard and embrace me.

4. He will give you information about the future.

This prophecy of Jesus caused Tertullian and other Trinitarians to redefine the meaning of the Spirit of Truth. This is referenced in *John 16:12-14* (NKJV):

"I still have many things to say to you, but you cannot bear them now. However, when He, the Spirit of Truth, has come, He will guide you into all truth; for He will not speak on His own authority, but whatever He hears, He will speak; and He will tell you things to come. He will glorify Me, for He will take of what is Mine and declare it to you".

Despite all of the above efforts, even Paul could not theologically sell the idea that Jesus, the Son of Man, is 'God', let alone a 'Ghost' who is God the Almighty, to the real disciples of Jesus—those who knew Jesus more profoundly than Paul himself. As soon as Paul preached his new theology against the teaching of Jesus, they ejected him (excommunicated him) from amongst them. Please see it for yourself referenced in *Acts 9:26-30* (NKJV):

"And when Saul had come to Jerusalem, he tried to join the disciples; but they were all afraid of him and did not believe that he was a disciple. But Barnabas took him and brought him to the apostles. And he declared to them how he had seen the Lord on the road, and that He had spoken to him, and how he had preached boldly at Damascus in the name of Jesus. So he was with them at Jerusalem, coming in and going out. And he spoke boldly in the name of the Lord Jesus and disputed against the Hellenists, but they attempted to kill him. When the brethren found out, they brought him down to Caesarea and sent him out to Tarsus."

The immortal intelligence asks you: If the original disciples of Jesus of Nazareth—who profoundly connected with and knew Jesus, who learned directly from him and were his actual students—rejected the teachings of the likes of Paul and Tertullian based on Jesus as God in the Trinity, how is it that you are following Paul and his narrative creator, Tertullian? Are you claiming you knew Jesus more profoundly than his disciples, who chased Paul away and "wanted to kill him?"

'To be or not to be' moment for you, my dear reader, as Shakespeare so eloquently puts it. You can no longer pretend! It is time to talk to your soul. Whatever your decision is, it will be justified through either fairytale desires of 'living Under-Grace' or impartial facts. It is your freewill, It is your call for your soul my friend.

The masked multiplication of the Singular God into the 'Trinity' was not justified through the propagation and indoctrination of people in 'peace' during the early centuries of Christianity. So, the matter culminated in the 'iron-fist' enforceability of the tyrannical dictatorship of the Roman Empire. They held a series of councils, starting with the Council of Nicaea in 325 AD (After Christ), which conferred the legally binding 'Nicene Creed', which affirmed the divinity of Jesus' coequality in every aspect with God the Father. This was followed by the Council of Constantinople in 381 AD (After Christ), which saw the Holy Spirit incorporated within God (*Eloh/Elohim)* and Jesus, who had already been added in the previous council. Then came the Council of Ephesus in 431 AD (After Christ), which proclaimed Mary, the mother of Jesus, as the God-bearer. Finally, the Council of Chalcedon in 451 AD (After Christ) canonized and made the Trinitarian creed the official imperial decree enforceable across all territories under the empire.

The process above, which took nearly 500 years after Jesus, fundamentally aligned Christianity and Trinitarianism as a single unit. Hence, any dissenting 'heretical' strand would not be tolerated thereafter. There were no hostages or detainees for Unitarians— anyone found contravening the empire's law would be eliminated until there was one empire with one Trinitarian Christian religion.

What happened to the Unitarian Christians? Unfortunately, they fled from the powerful grip of the empire and were persecuted over the years. Their religious beliefs became concealed due to the ongoing discrimination. A few would later convert to Islam after some 600 years AD, owing to the proximity of the teachings of the Unitarian God by both Jesus and Muhammad, who started teaching the same theology of the Unitarian God as well some 600 years later, this is accompanied by the fact that Muhammad held Jesus in high regard, so much so that two chapters of the Koran are named after Jesus' parents: Chapter 3, Imran (the grandfather of Jesus), and Chapter 19, Mary (the mother of Jesus). Both chapters contain the detailed story of Jesus' miraculous birth, message, and mission—ironically fulfilling the prophecy of Jesus that caused dreading Tertullian and other Trinitarians.

A primary example of early Unitarians includes the renowned fascinating disciple and companion of Muhammad by the name of Salman Al-Farsi, 'from slave to king'. He narrated his own biography, in which he details his transition between one surreptitious church to another in the remote villages within the periphery of the Roman Empire, which by his account, he suggests that Unitarians survived for hundreds of years after Jesus, but they barely persisted due to the consistent oppression and elimination they faced. Salman, who converted from Zoroastrianism to Christianity, recorded his last conversation with his pastor before he passed away, telling him: "Oh son, I do not know anybody else who is upon our theology, but do not despair. A prophet would be risen in the Middle East to revive the way of the prophets..."

In light of the overwhelming evidence presented above, it is clear that, according to Jesus' own definition, the doctrine of the Trinity has no basis or merit in Jesus' religion. If, therefore, Christianity is his religion, then Christianity should conform to

Jesus' teachings and repeal this odd doctrine. It is also an interesting fact that the Trinity never existed with the Biblical prophets who preceded Jesus in the Bible, such as Moses, Abraham, and Noah, nor with the first man, Adam, who also did not have natural parents.

On the primary evidence, it has also been proven above that the Bible contains some serious scientific discrepancies and errors, which disqualify the claim that this Book is the precise words of the God of the Universe in totality, the 'inerrant' verbatim voice of God, which ultimately is its downfall. Even its secondary or supporting evidence contain clear flaws in its coherency. Jesus, in explicit terms, gave his message from his own mouth, stating that no one can take away from him what he declared, such as the fact that he is a prophet.

Unknown authors, on the other hand, gave unsubstantiated messages alongside Jesus, such as Bible authors Matthew, Mark, Luke, and John. On the other side of the spectrum, a man known to be the enemy of early Nazarene Christians, Saul of Tarsus AKA 'St Paul', completely coined a new theology with a new covenant wrapped in the baseless but attractive gift covering known as 'Under-Grace', while in reality, there is no substantive prize inside the alluring gift package. He further promoted himself to be an Apostle by his own dream and wrote the bulk of what are the current Gospels, amounting to 14 out of the 27 books of the New Testament, shaping his own doctrine.

For all the rest of the books, which, by historical fact, were written after Paul's books, meaning, the influence of Pauline doctrine has been documented in the rest of the books as it is evident.

The immortal intelligence leaves it to you to reach your own conclusion about the Bible, given all the information and data that have been compiled above, with irrefutable references and facts. The immortal intelligence leaves with the repetitive statement: My friend, it has always been your personal free will that can lead your soul to the truth. **'It is you call for your soul!'**

The Immortal Intelligence Engages with Islam, Muhammad and the Koran

Islam, being the youngest of the Abrahamic religions, Muhammad began to preach the strict worship of the monotheistic 'Singular' God of Allah/Eloh in Makkah ('Mecca') Arabia, which aligns with the Singularity we are after. This occurred nearly six hundred years after the historically recorded unified and concocted effort to eliminate Jesus from Jerusalem and the Holy Land as a whole took place, a campaign led by the Jewish religious supreme authoritative council known at the time as the 'Sanhedrin', with the 'High Priests' at its head. They were in control of the second Temple of Jerusalem at the time, which precipitated the wrath of God, with Jesus declaring that this 'house', in which the prophets had been harmed, would perish from the face of the earth, with no brick left standing. This prophecy was fulfilled 70 years later at the hands of the Romans, as referenced in Matthew 24:1-2.

The Koran has something 'interesting' to say to those people who wonder why there is suffering on planet Earth. Is this the best existence God can create? The Koran says to those people you misunderstood the point of your existence. The purpose Allah brought you into this earth is to ask you the same question as you are wondering yourselves 'vice-versa'; do you deserve the best Allah can create? If so, then prove it! What did you think Paradise, the 'Garden of Eden' is? It is precisely the best world that you are wondering about—life without death, life without anger and dispute, without ageing, hunger, or sickness, with delights and enjoyments beyond the limits of our human brain's imagination. This is referenced in the authentic Hadith (Bukhari 3244): **"Allah said, I have prepared for My pious servants things which have never been seen by an eye, or heard by an ear, or imagined by a human being. " If you wish, you can recite this verse from the Holy Koran: "No soul knows what is kept hidden for them, of joy as a reward for what they used to do."**

Is it not amount to a fantastic contradiction 'one shoots him/herself in the foot' to be marveled and give standing ovation when President John F. Kennedy says to people in his inaugural speech in January 1961, "Ask not what your country can do for you, but ask what you can do for your country" who cannot give firm promise of anything in return, as he could not promise to himself a protection from untimely death, but yet when Allah says ask not what perfect existence Allah can give in this world, but ask yourselves what can I do of good deeds and efforts to deserve the best existance Allah will offer the next world, we are contemptuous, scornful or even sarcastic!

And further, when you company CEO asks you what asset and value have you offered to the company to deserve the promotion you are asking for? Whilst this appears not only acceptable but fair to your intellectual reasoning when it comes from your ordinary employer, yet it suddenly would become outrages from God = not to mention the asking of God is far greater than the request of mere promotion from your company boss.

Consequently, for that reason, the Koran is asking you and me, "What do you have to show to deserve that?" This is referenced in the Koran 29:2-3: **"Do people think once they say, 'We believe, ' that they will be left without being put to the test? We certainly tested those before them. And, in this way, Allah will clearly distinguish between those who are truthful and those who are liars."**

This is a stark contrast to the Pauline doctrine of 'Under-Grace', which demotes the Singularity of Allah/Eloh to one of many, one in three to be exact, in the concept of the Trinity as it only requires abstract 'dogmatic' belief. Further, the Koran strengthens Jesus Christ's position, which was to keep and uphold the commandments until the heavens and the earth disappear. In other words for as long as you live. This is referenced in Matthew 5:18: "For assuredly, I say to you, till heaven and earth pass away, one jot or one title will by no means pass from the law..."

To the average non-Muslim, however, Islam is the least popular religion because it is the least attractive. There is no 'free ticket' to paradise here, and there is no too-good-to-be-true fantasy of 'Under-Grace' here, as seen in some other religious doctrines. There is the absolute requirement of actions and efforts, which will play a pivotal role in your success—resembling 'a lot like' your everyday real-life effort requirements in attaining successful skills and education, building successful businesses and careers, buying and maintaining to pay rent or mortgage throughout your lifetime on Earth etcetera. Further, this resembles what Jesus preached about observing all requirements of worship in the form of the commandments of Elohim/Eloh, referenced in Matthew 5:17-20, before Paul altered his message and introduced a 'new covenant' while he claimed this law was fulfilled by the 'crucifixion of Jesus', according to Paul.

Noticeably, this is a meritless claim, as evidenced in Jesus' speech coded above. Jesus says in the same passage that this law of Eloh/Allah will remain active and applicable until the end of the world, referenced in Matthew 5:18: "For truly I tell you, until heaven and earth disappear, not the smallest letter, not the least stroke of a pen, will by any means disappear...."

The accomplishment of observing the law as Jesus envisioned is alive and vibrant in the Koran because the Koran is not a book that readily gives way to human interventions. Due to its construction of oral and recital memorization verse by verse by a large number of the Muslim population, a concept known in the Semitic Arabic language as 'Hafiz' which linguistically translates to mean 'guardians', though in this context, it means 'memorizers' as the two meanings overlap, it surpasses tens of millions of people without exaggeration. The Koran's fortification from corruption is so documented that even its least admirers admit to that fact historically. In our era, Western academics such as Nicolai Sinai, Fred Donner as well as Angelika Neuwirth, the renowned religious studies professor from the University of Berlin, all came to the same conclusion: the preservation of the Koran reigns supreme over any other ancient manuscripts from historical and philological standpoints. Hence, any disregard towards it must be based on one's subjective taste rather than its substance or preservation, such as those who disdain to comply with the commandments of God has solely the choice to reject as surprisingly conferred by the Koran itself referenced in 2:256 but not the choice to interject manmade doctrine or altering its commandments because the masses must hear the call of God in its purest and uncontaminated form.

To mount on top of the previous inconveniences but truthful blocks of obligations in Islamic life, the religion, to some degree, is 'perceived' to be synonymous with 'social death' due to its restricted prohibitions, notably known as 'Haram'. To many, almost everything that amounts to the underlying dopamine 'fuel' of enjoyment in life is prohibited, such as the enjoyment of beverages of intoxication such as wine and the like, including all recreational drugs. There are common misconceptions equating violations of these 'lifestyle' prohibitions to the breach of the covenant of Islam itself, which refers to your sincere, devoted acknowledgement to the singular source behind the universe. How unfortunate that misunderstanding is!

The reality is that a full breach of Islam only contravenes after breaking the first commandment in Islam, which is the concept of the 'Shahada'. In other words, if you turn your worship to any 'object'—without exception—other than the creator of the universe,

Allah/Elohim, a concept known as 'Shirk', which is a partnership in its direct translation, the multiplication of deities worthy of your worship, then Allah says that His forgiveness is strictly precluded for those who give their worship to any other deity, re-echoing the 'first commandment' of the law of Torah. However, He may pardon anything else based on His discretion. Many people misunderstand this concept; however, the 'Shirk', which is worshipping the wrong god—including the 'Selfanists', those who worship themselves here — means those who die in such a state, which allows redemption at any time before death.

Which means a faulty individual in this realm can correct his/her-self anytime before death approaches, at which time the test is concluded. Why redemption is not accepted at this stage of 'death' because the veil is removed from the eyes in this actuality 'definitive' moment? Every human being is capable of seeing what was the unseen world to him prior to this juncture, what was dark energy and matter ascertained through 'impactism' in a vivid sightful dimension henceforth everyone would attest to the Singularity, a pardonship granted at this point would amount to passing the exam on a complicit cheating resources. This is referenced in Koran 50:22: **"It will be said to the denier, "You were totally heedless of this. Now We have lifted this veil of yours, so Today your sight is sharp!"**

Many of Muhammad's prominent disciples, known as 'companions', including the successor of Muhammad, Umar bin

Khattab, were pagan idolatry worshippers themselves before they converted which proves the above point. This is referenced in the Koran 4:116: **"Indeed, Allah does not forgive association with Him, but He forgives what is less than that for whom He wills. And he who associates others with Allah has certainly gone far astray."**

Beyond its antagonistic approach towards the drinking culture and the consequential parties that come with it, Islam also forbids eating pork (pigs), whom the Koran alludes to as having once been humans or creatures with free-will capabilities who incurred the curse of their Lord. This is referenced in 5:60: **"Say, ˹O Prophet, ˹"Shall I inform you of those who deserve a worse punishment from Allah ˹than the rebellious˹? It is those who earned Allah's condemnation and displeasure—some being reduced to apes and pigs and slaves to fake gods. These are far worse in rank and farther astray from the Right Way.""**

Hence, the reason why pigs and apes 'scientifically' carry the utmost similarities to homo sapiens, ranking as the highest animals 'per capita' who share similar DNA, estimated to be 98% with humans. The fact that 'in some respect', we may have even closer DNA with pigs than apes is why many major medical researchers use pigs for their experiments. This spans from metabolic processes to body proportions, as published in *Pig Genomics and Genetics* by the National Institute of Health (NIH). This places question mark on the accuracy of the theory of evolution, which claims humans have exclusive ancestors in apes known as 'Hominins'. This sounds more beautiful and dignified than admitting that humans may also have an ancestral relationship with pigs. While this is inferred as a speculative deduction on my part, in the meantime, if we were to admit this scientific fact, pig's meat being our favorite dish in the West would equate us with cannibals and, therefore, cannot be morally palatable! But the good thing is DNA evidence does not 'politicize' the facts, unlike our national countries' news channels. This is why each major country made their own news channel in order —to create a narrative that is more palatable to their liking and subjective views. What is even more astonishing is that the Koran informed us of this accurate scientific information more than 1,400 years ago (as of 2024), and yes, the Koran is that good.

The Koran further prohibits any animal slaughtered without the appreciation of its creator via declaring 'His' name when commencing the butchering of such an animal that you did not create yourselves. Thus, Islam reminds you and me, the next time we eat our fast-food 'burger', who permitted you to kill such an animal? In other words, eating animal meat can only be justified through the consent of its creator. Which, in turn, aims to relieve the guilty conscientious burden of vegetarians, even if they disagree with God. Islam is drawing our cognitive intelligence towards our actions; it holds the human species in high esteem for moral and thinking accountability, something we ourselves testify to, as we called our species 'Homo Sapiens', which we gave the meaning of 'the thinking person', concurring with the factual deliberations above.

With all the above restrictions, Islam further requires upon the individual daily rituals of worship to the God of the universe, perhaps more than any other contemporary religion. Five times a day, the Muslim person is obliged to join with the Lord via offering their prayer. There is no mild worship of music and dance occurring only on weekly intervals in church/temple services here. The interconnection between the servant and the Lord continues in an unbroken chain; it is an exclusive personal affair between the two parties. The Koran makes it clear that this special bonding relationship between God and His servants must be a direct interconnection, with no parties in the middle—no intermediaries at all: not a prophet, not an angel, nothing whatsoever can place themselves in between the human and his/her Lord. Astonishingly, 'the ultimate freedom' you did not expect lies in this special relationship. The Koran highlights the paramount importance of this and, therefore, prescribes to its followers to pray to God so that they can attain the 'straight path' to connect to God on a daily basis, at least no less than seventeen times, as referenced in Koran 1:6: **"Guide us upon the Straight Path."**

In fact, contrary to common misunderstanding, the Koran calls this a privilege and favor conferred upon mankind, for which their intellect has become recklessly foolish not to take advantage of this great offering from the Creator of the universe as they turn their

277

worship to merely symbolic names with no Godly powers, referenced in Koran 12:39-40 which is better too many different gods with conflicting demands or the only Supreme Allah, these intermediaries whom you give the right to worship only reserved for Allah have no merit: "….**Which is far better: many different lords or Allah—the One, the Supreme? Whatever ˹idols˺ you worship instead of Him are mere names which you and your forefathers have made up—a practice Allah has never authorized. It is only Allah Who decides. He has commanded that you worship none but Him. That is the straight path, but most people do not [care] to know."**

Once upon a time, an argument ensued between a devil's advocate and a God's advocate. The devil's advocate said, "The devil has powers. "Yes, said the God's advocate in reply, "So does my local Mayor; he can give you free garbage collection from your door for a year, but neither can make you any younger nor shield you from death."

The bucket of the to-do list is further skin-deep, as it includes the well-renowned month of Ramadan, during which Muslims are mandated to carry out fasting from twilight to twilight, dawn to dusk. There is no teasing or gimmicking here, as some cultures only fast from protein, meat, dairy, and other animal-related products etcetera. Fasting in Islam is an organic partaking; in other words, its scope applies to all food, solid and liquid, including bread, vegetables, and water. Of course, anomalies are granted for the vulnerable class of society, such as the elderly, children, the ill, and those who are travelling, among others, who are excused. To many non-Muslims, this seems an insurmountable task, and thereafter, they claim that this is not possible, that it is a 'slow-motion' death sentence.

They could not be more wrong, evidently, not only there are a large community of people who are created from the same substances as 'themselves' are observing the fasting from sunrise to sunset, but they are thriving. If anything, the number of the global Muslim population is increasing and is expected to reach over 30% of the world's population, up from 25%, by the year 2060, according to PEW Research. If Ramadan were as harmful 'killer virus' as many fear, the evidence would have been reversed, and the Islamophobic media would have reported like piranha fish feasting—'Islamists' shot themselves in the foot by committing mass suicide due to their self-righteous fasting during Ramadan, instead of 'desperately' calling 'terrorism' an Islamic act rather than its objective description of a criminal act.

This religion is somewhat a self-opting prison to its critics. They render their personal verdicts and say to 'themselves' where is the happiness of life if you cannot drink wine or beer, eat barbecued pork chops on summer holidays, or participate in dating without fulfilling the spousal requirement of upgrading your partner to be a legally recognized wife or husband—so-called 'situationships or friends with benefits' these days as they used to call these informal unions 'courtship' partnerships or 'concubinage' relationships in medieval times. Nothing is inherently 'modern' about this; human behaviour is such that history is bound to repeat itself. So, they frantically ponder and wonder with the conjecture of deflecting the underlying veracity: which religion preserves the Singularity of God in the most pure and organic way possible? Is Islam a serious candidate, perhaps?

Have you ever heard the expression 'I don't want to know mode'? An internal heart versus brain-battle ensues. It is the thought one traps him/herself with when they realize upon investigating the matter further may lead to 'truthful' but undesired outcome. The immortal intelligence, in a paradoxical 'abstract sense', appreciates such a standpoint. Even the wife of Muhammad, Aisha, appreciates this 'natural' human feeling, as she said [paraphrased] if Islam was all about don'ts and 'Haram', it would have lost its justifiable appeal to people, suggesting it is much more than that, referenced in Authentic Bukhari 4993: "......If the first thing to be revealed was: 'Do not drink alcoholic drinks, ' people would have said, 'We will never leave alcoholic drinks, ' and if there had been revealed, 'Do not commit illegal sexual intercourse, ' they would have said, 'We will never give up illegal sexual intercourse......" This notion is something the immortal intelligence will engage with further in our conversation.

In all realities, who wants to hear more obligations, a cluster of rulings, more don'ts coupled with more consistent rituals of limbs to prove what the mouth proclaims? My dear reader, if you told your spouse you loved them, 'you better prove it'—a question many of you are intimately familiar with from the other half in the 'kitchen sink' debate 'reminders' from time to time.

For instance, who wants to come to know the prices went up after the New Year, and your monthly obligatory bills are higher? But the truth is, it is not the desires that lead us to our actions; rather, it is the essential necessity that sets our priority and overrides those juvenile resentments. This helps you protect yourself from getting caught at the wrong time later. Remember, the defensive mechanism of immortal intelligence is activated at this juncture to override. If one neglects to pay bills early on, he/she will deal with the bills, along with surcharges of late fees and County Court fees at a later stage, with dire financial consequences. Most of you know this fact and address your worldly obligations, whether you desire them or not.

But the inevitable purpose of life can be compromised is one's foolish way of dancing to fall into the unknown reality of the grave. What is the alternative? The unknown reality of cremation or the myth that we shall not feel any pain after death? Have you ever felt pain 'nightmare' in your dream that people around you were completely unaware of your predicament? The immortal intelligence surprisingly points the anxious hearts, after all, may change after trying new 'rituals' as they always do. Perhaps all you need to do is step over the fence once and realize there is no inferno on the other side.

For those of you who fear Islam is the inevitable Titanic ship that will sink and therefore bring a tragic sad end for those who boarded my dear reader, you could not be far more mistaken than those who were astounded by the size and stature of the Titanic ship, with its short-lived glamour and thin-layered gleam of shine and light. They thought to themselves, this ship is magnificently 'unsinkable' with the foolish utterance of "Even God cannot sink it!" Conversely, Islam claims it has Pandora's Box of happiness, containing inside of it full of stretchy delights, from springy candies to Jackfruits that seem to entertain forever due to their elasticity—of course, in a metaphorical sense.

The Koran, on the contrary, challenges our understanding of what is happiness and makes a very bold statement that actually claims profound 'intrinsic' happiness can only be attained through joining with Allah, referenced in Koran 13:48: "**.....Surely [only] in the remembrance of Allah do hearts find comfort. "** The Koran is demonstratively challenging our understanding of happiness—or lack thereof—what constitutes it, the ultimate prize in life. The reason why you spend all of the money you earn can broadly be summed up as seeking happiness—searching for that joy and contentment. If you were to calculate what percentage of your expenditure is consumed on this topic of searching for happiness, chances are it would command the majority, the better part of your income, for the average reasonable person. Be it evenings out, weekends, or holidays, the bottom line is finding pleasure costs us repetitively, as we know for a fact.

However, our conversation would remain a fairytale and fiction or even somewhat deceiving as the artificial orbit sells 'the party culture' unless the immortal intelligence briefly engages with the genuine dark reality of the brain's intoxication caused by all recreational drugs and drinking. As the name suggests, the natural neurotransmitters of the brain are artificially interfered with for the purpose of inducing non-organic happiness out of the brain. In other words, the brain is 'momentarily' out of touch with reality, perhaps partially lunatic, or in some cases, when it reaches its limit, it becomes fully lunatic. The ways this is achieved are through all intoxicants, without exception, including alcoholic beverages. These substances 'artificially'—without cause and effect—increase the neurotransmitter known as dopamine, which 'otherwise' plays a crucial role in achieving happiness for humans. Hence, in lieu, happiness should be emanating from real accomplishments, together with the encouragement to augment the accomplishments already attained in order for the person to reach his/her optimum level of success, rather artificial methods provide fake happiness.

An example is conferred: It is a lot like the mountain climbing sport. As the climber reaches new heights, dopamine is released to give him/her not only meaningful happiness due to the achievement reached, but the dopamine also serves as the motivating catalyst that encourages the climber to go further and further until he/she reaches the ultimate height of success. Hence, creating a habit of success, which Allah intended for, the Koran repeats this notion again and again: 'nonstop' striving to do good deeds of justice and fairness, charity, and compassion—not only towards your fellow man but towards the environment that is under your control as well—in order to score 'your personal best', referenced in amongst many verses in the Koran, including 49:13: **"O mankind, indeed We have created you from male and female and made you tribes and nations that you may recognize each other [treat each other with fairness and equality]. Indeed, the most noble of you in the sight of Allah is the most righteous [the one who strives to become the most righteous]. Indeed, Allah is Knowing and Aware [what you do]."**

Based on the non-infiltrated, naturally occurring dopamine resulting from real achievements, which in turn induces the happiness and motivation to go on and reach higher personal growth as a recurring habit, the immortal intelligence will call it 'the healthy addiction'. The habit of being addicted to consistent success is a variant of addiction as well, which the above Koran envisions for mankind.

However, the induced happiness, which comes as the result of brain infiltration from intoxicants such as alcohol and drugs, leads to the opposite reaction. This is because the dopamine is the result of the drinking/drugs 'itself', an illusionary achievement. It is a deceptive trick conducted on the brain, which is why the admittance of 'let us get wasted' becomes a self-fulfilling prophecy. Not only is one wasting a portion of his/her time, but also a portion of their life, along with it, a portion of their brain, not to mention a portion of their hard-earned money. The way it impacts the brain is that instead of yearning to achieve more from dopamine resulting from reality 'authentic achievements', your body would long for the feeling under the false induced dopamine. This, in turn, creates unhealthy addiction, as the increased dopamine wears off—otherwise known as 'withdrawal symptoms'. The poor alcoholic would return to drinking more alcoholic beverages. Perhaps alcoholics and addicts would have been workaholics and dedicated achievers had their body adapted to the reality of the 'healthy addiction' rather than the illusionary happiness of intoxication.

With that said, the Koran makes an unusual remark exclusively relating to alcoholic intoxication. It underscores some benefits that can be experienced from it, as referenced in Koran 2:219. Drinking has some benefits, but the harm outweighs the benefits: **"They ask you about wine and gambling. Say, 'In them is great harm and [yet, some] benefit for people. But their harm is greater than their benefit."** The Koran is referring, in terms of the benefits with respect to alcohol intoxication, specifically, the fact that a small quantity of alcoholic intoxicant is capable to improving Gamma-Aminobutyric Acid, which in turn induces a sense of calmness and relaxation from the brain and regulates anxiety and stress-related downturns. However, this is where the good news ends. Add a bit more than a small quantity into the body, sedation and mental impairment will take over, as precisely the Koran suggests: the harm will quickly outpace the benefit.

Further, according to the Koran, nothing weighs more valuable as an asset than the human brain, referenced in Koran 30:8: why does mankind not apply the intelligence We gave them about themselves and realize the creation of the heavens and earth cannot be without a purpose, and derived from a singular source? They are facing death to return to their Lord, but most of them are ignoring the fact of their appointment with their Lord **"Have they not reflected upon their own being? Allah only created the heavens and the earth and everything in between for a purpose and an appointed term. Yet most people are truly in denial of going back to their Lord!"**

Hence, the brain is connected to thinking and reflecting on my and your purpose of life. Furthermore, our attention is drawn not only to our own death at some point—hopefully in the distant future—but to the 'blanket' universality of death covering everything we know. Even the planets and the sun are also facing death, referenced in the above verse, 'appointed term', as science itself concurs with the above verse. The sun is projected to deplete power to sustain its life, a process known as 'The End of Hydrogen Fusion'. Yes, the Koran is that good. Further, the Koran uses a unique adjective to describe when it comes to the process of death. It says in 21:35 that every soul will taste death. As opposed to simply saying we will die, it is telling us that the process 'in and of itself' is an experience, just like the food you tasted. It has a flavor and feel, either good or bad, and everything in between, with the franticness of its nature being alien to us—those of us who are living. For amusement's sake, let's measure it as though we are reading this book, for I do not know of any ghosts reading books. There is, though, the hope for calmness with the description of tasting food within the frantic and uncertainty associated with its unknown nature, that is, there is that possibility of a good sensation that comes with tasting good cuisine. Hence, since the process of death has been compared with tasting food, I pray to God that our death be like visiting our favorite restaurant.

Koran 21:35:**"Every soul will taste death. And We test you, ˹O humanity,˺ with good and evil as a trial, then to Us you will ˹all˺ be returned."** If you ever wanted to know why are you created, why go through life and death, you find it in the above verse: in order for the Almighty to try and test us with good, bad, and everything in between.

But there is an even more primary reason than the above as to why we have been created, with a unique verse exclusively about the Singularity of God, pronounced in explicit terms, referenced in Koran 51:56: **"I did not create jinn and humans except to worship Me."** Yes, everything in existence was conceived for this reason alone, that includes people and jinn as well, including Satan himself. The universality of death is so rigid blanket, so much so that nothing can escape from the scope of its effect: not a human, not jinn and Satan, not an angel, not planets and stars, not even prophets. Consider this verse: Allah tells His beloved prophet Muhammad, if you cannot escape death, does anyone else stand a chance? Koran 21:34: **"We have not granted immortality to any human before you, ˹O Prophet˺. So if you die, will they live forever?"**Due to this understanding, it is foolishness to expect to miss the accountability before Allah. Indeed, for what other reason are we going to taste death otherwise?

For the human brain to have clarity of reflection and thinking to contemplate and ponder about themselves and their ends, therefore, intoxication, which gravitates the mind in the opposite direction of playfulness, sedation, and, more crucially, impairment of reasoning, particularly the prefrontal cortex. Furthermore, the habit can, in the long term, lead to brain decline altogether. As such, the Koran has found it unsuitable for Adaminians 'the children of Adam', who are greatly anticipated beings with the greater aim to decipher the purpose of life, to be wasted so willingly and happily and say, 'let us get wasted', Rather, no, let us get preserved ought to be the phrase, as envisioned by the Koran. But the freedom to be wasted has been granted to who so ever wishes to do so.

Many people wonder: Is Allah a ruthless sadist, and why does He intend to take away all the pleasures and delights of life? Usually, these types of complainants have in mind recreational drugs, drinking, intimacy outside marriage, and perhaps as far as even usury, which means trading loans with the expectation of interest in return as profit. This is further compounded by many serious acts of worship, such as praying five times a day, fasting all daylight during the month of Ramadan, and the like.

The Koran addresses this feeling of certain people on the back of the verse which prescribes fasting as an obligation to all believers wherever they are, referenced in Koran 2:185: **"Ramadan is the month in which the Koran was revealed as a guide for humanity, with clear proofs of guidance and the decisive authority. So, whoever is present this month, let them fast. But whoever is ill or on a journey, then ˹let them fast˺ an equal number of days ˹after Ramadan˺. Allah intends ease for you, not hardship, so that you may complete the prescribed period and proclaim the greatness of Allah for guiding you, and perhaps you will be grateful."**

We are particularly interested in the part of the passage where it says: **"Allah intends ease for you, not hardship so that you...[later] be grateful. "** It is a conundrum, as in the earlier part of the passage, humans have been asked to fast during daylight, but the latter part of the same passage tells us this is supposed to make your journey easier, not difficult. So, what is going on? How can fasting be effortlessly easy?

In order to understand what will make us grateful or appreciative later, we have to understand the purpose of these disciplines of worship because that is what they are, after all—a routine prayer, a routine fast every year for a month, a routine paying of obligatory charity of 2.5% on your income for those who qualify etcetera.

The most decorated Olympian in history, Michael Phelps, was given a disciplinary routine to follow in order that he shall accomplish to become a 'grateful' champion later, by his longtime coach Bob Bowman. Make no mistake, the coach was very well aware of how potential Michael is as a player before he even competed, but the choice to grow and reach that triumph was ultimately up to Michael with his free will. Consequently, coach Bob gave his player the winning strategy of a tight disciplinary routine if he sincerely wanted to compete at the highest level and even win the highest prize—the gold medal. This training included, at its peak, 50 miles of swimming every week, and he was required to carry out this task twice every day for 6 days a week.

Both Michael and Bob understood these measures must be taken in order to compete at the highest level of the championship. Additionally, there are many more gold medalists, such as Mo Farah, a long-distance runner. Mo and his coach, Alberto Salazar, knew the kind of effort they had to invest in the endeavoring journey to compete at the highest level and become winners. Hence, Mo was prescribed to train by running 135 miles every week, training two sessions a day with no rest day except on Sunday, when he performed one session of 20 miles. If this is the information revealed, the immortal intelligence projects that there were extra secret training disciplines these athletes partook in so that they could attain a competitive advantage over their rivals.

The bottom-line truth is that none of these athletes questioned whether their coaches were sadists or hated them. None of them questioned why they were doing these 'tough' routine disciplines because they were true to their goals, and above all, they understood what it takes to be winners.

However, you might say, "Yes, the above is true for those athletes, but I did not opt to become an Olympian, and I did not sign up for a high-level sports competition. What does that have to do with my ordinary life? Can I not just relax, 'chill', the all-too-common mindset?" My dear friend, according to the Koran, yes, actually you signed up to a test called 'life' itself. And further, yes, you signed up to become the privileged human being you are, with the license to eat well-done steak or medium-rare, whatever your preference, or lobster that you did not create for yourself, or the dessert you enjoy, such as Tiramisu, with God-created ingredients from eggs, dairy-derived cheese, grapes, sugar, and cocoa powder. Is it not an overlooked fact that we humans, as arrogant and boastful as we are, cannot create one 'natural' ingredient for ourselves?

The immortal intelligence finds this clear, inferable fact from the Koran, which informs us that, yes, 'natural selection' took place. Not the purposeless, aimless 'natural selection' endorsed by the artificial orbit of Charles Darwin, but one with a higher aim took place. This explains two things: why are we ageing and facing inevitable death, and why do human species command higher intelligence quotient (IQ) and aptitude than other species on the planet Earth? Hence, in the Koranic 'natural selection', all creatures were asked: who is going to opt to become a free-will capable, highly intellectual species and consequently would be liable for punishment should they waste their intelligence with the recognition to acknowledge and thank the unseen singular Creator? On the contrary, those who acknowledge the fact that they are created and cease pretending to be in control before their freedom of life is revoked are going to be rewarded. Thus, whoever signs up to become a free-will capable creature, such as a human being, their memory of this divine contractual agreement 'deal' during the test will be removed from their 'remembrance' in the hippocampus part of the brain, only to regain memory of it upon concluding the test at the stage of death, at which time the test ends. This is referenced in Koran 33:72: "**Indeed, We offered the trust to the heavens and the earth and the mountains, but they ˹all˺ declined to bear it, being fearful of it. But humanity assumed it, ˹for˺ they are truly wrongful ˹to themselves˺ and ignorant ˹of the consequences˺.**"

If you are amongst the complainers and haters of this undertaking, look at what the verse describes you as—it was a grave mistake on your part to opt for it. Hence, you wronged yourself and are ignorant about the consequence, which is 'all-time' jail, as opposed to the maximum criminal penalty of a lifetime in our world. Because in the afterlife, there is no death, hence any punitive sentence ('jail terms') will feel like forever. In fact, let us engage with what the scale of time is in the next universe, promised by the one who created this ever-expanding first universe.

Before that, however, the Koran also gives away the contractual agreement of this test, the memory of which has been removed from our minds, in what is known as the 'Humanity Covenant', referenced in Koran 7:172-173. This allows us to view the agreement, which our immediate recollection cannot grasp in the duration of the test while it is ongoing. It is recorded in two major clauses.

In clause one: we reaffirmed that our God is only the singularity Allah/Eloe/Elohim, however dialect and accent you may pronounce it, or whatever substantial description you may acknowledge Him as—for instance, the Supreme Singular God as seen in the Hindu scriptures and the like. The reason we reaffirmed this is extracted from the nature of the question: "Am I not your Lord?" This demonstrates the fact that we had previous knowledge of the status of our Creator, according to the verse. Hence, we infer that this clause is a 'certification statement', not a novel clause in the relationship between the two contractual parties—'God and creatures'.

Clause number two: is a type of 'liability clause', where failure occurs on the part of the children of Adam to uphold their obligations within the contract gives rise to a 'criminal' punitive sentence to be carried out against them. It further clearly states the inadmissibility of transferring or inheriting responsibility ('sins') to anyone else, such as parents, tradition, or culture, as it bars the erection of such rights and claims. It clearly forewarns such scenarios: [paraphrased] do not later claim that 'the breach of the covenant' was a tradition we followed from our parents. Please see the bare-bones contract in the Koran, referenced in 7:172-173:

"And ˹remember˺ when your Lord brought forth from the loins of the children of Adam, their descendants and had them testify regarding themselves. ˹Allah asked, ˺ "Am I not your Lord?" They replied, "Yes, You are! We testify. "˹He cautioned,˺ "Now you have no right to say on Judgment Day, 'We were not aware of this.' Nor say, 'It was our forefathers who had associated others ˹with Allah in worship˺ and we, as their descendants, followed in their footsteps. Will you then destroy us for the falsehood they invented?'"

The time scale is completely different in the promised new universe. According to the Koran, everything from time to activities has been enlarged so expansively that what we call a lifetime in this universe will amount to an hour in the next life, referenced in Koran 10:45: **"On the Day He will gather them, it will be as if they had not stayed ˹in the world˺ except for an hour of a day, ˹as though they were only˺ getting to know one another. Lost indeed will be those who denied the meeting with Allah and were not ˹rightly˺ guided!"**

The Koran tells us that the Judgment Day alone is as far as fifty thousand years of what we count in this world, referenced in Koran 70:4:˹ **"through which˺ the angels and the ˹holy spirit will ascend to**

Him on a Day fifty thousand years in length." The 'holy spirit' is the title given to the special Angel Gabriel in Koranic terminology, but it is not limited to him alone. Let us face it, because Allah is the creator of many spirits and souls, many of them are holy. Furthermore, any subsequent day from that day forward, whether in enjoyment or punishment, will amount to a thousand years of what we know as time in this universe, referenced in Koran 22:47:

"They challenge you, Oh Prophet' to hasten the torment. And Allah will never fail in His promise. But a day with your Lord is indeed like a thousand years by your counting."

However, in the next universe, where it has been narrated from Muhammad in authentic Hadith Muslim 2837 that no Sun is orbiting any longer and there is no ageing at all, then in such a scenario actually time is therefore 'stagnant' and no longer progressing on a 'vertical axis', while we are, on the other hand, moving and progressing—a paradoxical 'conundrum' in our solar/lunar based time as the immortal intelligence finds. Therefore, in all respects, the Koranic illustrations of a new universe with the absence of solar and lunar rotating and moving in their axis make the time infinitely longer than our current system. This further augments our comprehension of how important the next universe is.

For that reason, your immortal intelligence should give higher priority to that which is infinitely everlasting and longer. An example is given: in the fairly new world of electric cars with variant transmissions and gearing system hence, there is no rev counter.

Therefore, you explain to someone who has no clue and say to them that the acceleration in an e-car is going at a constant pace and does not rescind 'stop' on Revolving Per Minute (RPM) threshold or gear change because there is no traditional transmission involved at all, just like the absence of traditional Sun and Moon orbiting in a solar system, thus affecting time. Hence, time is perhaps on a motionless horizontal axis in this existence. Anyhow, according to the Koranic understanding, time will be under constant position in the next universe. Furthermore, in terms of light and energy, it seems every object in existence—including people—would generate their own energy, including their own light, which explains the reason why the Sun is no longer required, referenced in Hadith authentic Bukhari 6567-6568: "...if one of the women of Paradise looked at the earth, she would fill the whole space between them (the earth and the heaven) with light..."

For the impartial eye and the fairer minds to facts, the Koran has a unique universal appeal and inclusivity for people of all backgrounds beyond the negative smear and slander towards it, unlike other religious books. There is no legitimate major scripture that culminates all the people, with their diverse races and backgrounds, along with their prominent leaders, in one basket other than the Koran. Yes, that is true. If you are Buddhist and hold high esteem to apply for the teachings of Buddha, such as becoming humble and uplifting those who suffer from hunger and poverty, you would find the Koran home, referenced in Koran 25:63: "**The**

´true` servants of the Most Compassionate [Allah] are those who walk on the earth humbly, and when the foolish address them ´improperly`, they only respond with peace."

Additionally, uplifting poverty and hunger: the Koran made loans with interest ('usury') illegal so that the circumstances which creates poor people to consistently remain poorer and enslaved to a few rich elites are rectified in order that wealth distribution extends to the middle and lower classes of society. This is referenced in Koran 59:07 the state wealth should be extended to "..the orphans, the poor, and the needy travelers so that wealth may not merely circulate among your rich." In other words, there should be state policy where the distribution of wealth reaches the general population as a whole and does not remain in the hands of a few, something we are seeing more and more in this modern world—the disparity between the rich and the poor gaping significantly.

Furthermore, if you are a Hindu and want to connect to the supreme God as mentioned, 'Ekam Evadvitiam', He (God) is only one, without any addition—'one without second'—in the Chandogya Upanishad, Chapter 6, Section 2, Verse 1. You are also at home with the Koran, referenced in 2:21-22:

"O humanity! Worship your Lord, Who created you and those before you, so that you may become mindful ´of Him`. ´He is the One` Who has made the earth a place of settlement for you and the sky a canopy; and sends down rain from the sky, causing fruits to grow as a provision for you. So do not knowingly set up equals to Allah ˹in worship˺."

If you are of the Jewish race of Israeli descent and therfore love Moses with a sincere heart, you will find him in the Koran—his story, his mission, his struggles, and most importantly, his teachings—referenced in Koran 3:2-4:

"Allah! There is no God ˹worthy of worship˺ except Him—the Ever-Living, All-Sustaining. He has revealed to you ˹O, Prophet˺ the Book in truth, confirming what came before it, as He revealed the Torah and the Gospel."

Further, Allah sends Moses and Aaron to the tyrant Pharaoh to release the children of Israel from the house of bondage and slavery, referenced in Koran 20:47:

"So go to him and say, 'Indeed, we are messengers of your Lord, so send with us the Children of Israel and do not torment them. We have come to you with a sign from your Lord. And peace will be upon he who follows the guidance."

If you love Jesus Christ, you will most definitely find him in the Koran as well—with his mission, struggles, and message. He is directly mentioned more than Muhammad himself, twenty-five times. In fact, the Koran is the true fulfillment of Jesus' prophecy in John 16:12-14 (NKJV):

"I still have many things to say to you, but you cannot bear them now. However, when He, the Spirit of truth, has come, He will guide you into all truth; for He will not speak on His own authority, but whatever He hears He will speak; and He will tell you things to come. He will glorify Me, for He will take of what is Mine and declare it to you. "

As Jesus clarified the specific 4 signs of the Spirit of Truth:

1. He will make things that you cannot bear now clear to you.

2. He will not speak of Himself; therefore, He will speak what He hears—'Revelation'—from the Father Eloh/Allah.

3. He will speak of me in high regard and embrace me.

4. He will give you information about the future.

The Koran proves that itself and Muhammad are the fulfillment of the above prophecy. The immortal intelligence finds no other candidate who captures all the signs contained in the prophecy more so than the Koran and Muhammad.

First, the Koran calls itself a spirit, referenced in Koran 42:52:

"So We have revealed a spirit to you [Prophet] by Our command: you knew neither the Scripture nor the faith, but We made it a light, guiding with it whoever We will of Our servants. You give guidance to the straight path."

The angel who delivered the Koran to the Prophet Muhammad is called the Holy Spirit—Angel Gabriel, referenced in 16:102:

"Say, 'The holy spirit has brought it down from your Lord with the truth to reassure the believers and as a guide and good news 'gospel' for those who submit ˺to Allah˺."

As such, therefore, the Koran itself is the spirit of truth.

In addition, the first sign—the prophecy that the spirit of truth will explain things you cannot bear now— is stated clearly and without a doubt in the Koran, this is referenced in the Koran 12:111:

"...This message cannot be a fabrication, rather, it is a confirmation of previous revelation, a detailed explanation of all things, a guide, and a mercy for people who believe with certainty".

The second sign—He will not speak of his whim 'himself', He will speak what He hears from God—this is referenced in the Koran 53:2-5:

"Your fellow man (Muhammad) is neither misguided nor astray. Nor does he speak of his own whims 'imaginations'. It is only a revelation sent down to him. He has been taught by one ˈangelˈ of mighty power."

The third sign—Jesus would be revered in the Koran, referenced in 5:46:

"Then, in the footsteps of the prophets, We sent Jesus, son of Mary, confirming the Torah revealed before him. And We gave him the Gospel 'the good news' containing guidance and light and confirming what was revealed in the Torah—a guide and a lesson to the God-fearing."

The fourth sign—He will give information about the future. The Koran gives a large amount of future incidents, but the immortal intelligence has chosen this future incident for its relevancy regarding the subject matter. In the Judgment Day, the singular God Allah/Elohim asks Jesus a direct question: did you and your mother claim to people that you are gods besides Me, referenced in Koran 5:116:

"And ˹on Judgment Day˺ Allah will say, 'O Jesus, son of Mary! Did you ever ask the people to worship you and your mother as gods besides Allah?' He will answer, 'Glory be to You! How could I ever say what I had no right to say? If I had said such a thing, you would have certainly known it. You know what is ˹hidden˺ within me, but I do not know what is within You. Indeed, You ˹alone˺ are the Knower of all unseen.'"

This is one of the future incidents that we can prove now, which is the reason why there is no verse in the entire Bible—not even one in direct language—in which Jesus says, 'I am God the Almighty' or 'worship me'. Not ambiguous, not shadowy wording such as 'I am Alpha and Omega', 'I am', 'I and God are at one', or 'If you see me, you see the Father' and the like, which is not a clear claim of divinity for those who have an impartial, objective eye. For most certainly, Jesus could have clearly stated, in unambiguous terms, 'I am God; besides me, there is no God, ' the likes of the first commandment, but he did not.

Primary Evidence:
Science versus Koran

Despite all the impressive information detailed above, the immortal intelligence requires primary evidence. Although certain primary evidence has been glossed over in our previous discussions, we still require some clear proofs that justify the Koran and Muhammad's claim, that is, the Koran is revealed by God and Muhammad is his Prophet. The truth is, none of us can turn the clock back and pay a visit to the Prophet of Islam in order to gain first-hand experience with him. Therefore, we will examine what primary evidence the Koran puts forward 'brings to the game' to justify its validity; otherwise, all the amazing stories discussed above are at risk of amounting to just that—'an amazing story'.

To many, however, particularly in the advanced nations, have a reluctancy to even fathom that the Koran can be a scientific book. Specifically, Muhammad, its 'central figure', who delivered it to the world as an Arab man living in Arabia in the sixth century with a donkey named 'Dul-Dul' and his long-distance main vehicle being a camel named 'Qaswa'. If that is not enough of a turn-off, then some of them learn the fact that he has been known as an 'illiterate' man who could not read or write, referenced in Koran 7:157, residing in the desert of Arabia with no prior civilization. The mere logic dictates to them, why would I listen to this prehistoric, technically 'premodernity' claim of people using the name of God to gain power and influence over the ordinary folks, such as farmers and shepherds, in that harsh desert environment? It is not for 'me' in 21st century with all the scientific and technological advances, etcetera.

However, what they fail to realize is that this is the reason why Allah chose him from such an environment at such a time, with no scientific pedigree from nomadic 'Badwan' Arabia in the sixth century. The question that bewilders everyone is: you already know the above-described man in sixth-century Arabia should not 'by any measure' be the most scientifically minded person, with the largest and most accurate amount of scientific predictions produced by him. He had never conducted scientific research, never had access to a laboratory or any scientific equipment. He never had access to ultrasound technology to give precise details of pregnancy and child formation in the womb, nor advanced space telescopes to give accurate astronomical data regarding the universe in which we exist in its expansive interior parameters.

Then the Apollo 13 phrase, 'Houston we have a problem' knocks our heads. If mindboggling ever described something, this is it my dear reader—there is a colossal conundrum! It is clear that he cannot predict scientific theories that would be proven centuries and millennia later. No way— it is impossible! The only sensible conclusion is that, indeed, he is who he says he is—'a prophet of Allah'—and his information was a revelation from Allah, the Singular God. Otherwise, you have lost your mind and left it in the pub/bar or at the last holiday you went on for a measure of humor within a serious intellectual debate. Let us dive into the scientific reports in the Koran.

This book, called The Koran, is as daring as it is extremely confident in its abilities. So much so , that it calls its critics and foes, 'Here I am' on a silver platter—prove me wrong. Catch me 'at fault' if you can. The interesting reality about that is that it has been drawing our attention for over 14 centuries. Verses like this, If any other than the Singularity Creator Elohim/Allah had authored it, there would have been many errors and mistakes, as they are bound to happen for perpetual theories and predictions because any other research, scientific or not, occurs on the back of 'trial and error' that can be witnessed. However, there is none of that here in the Koran, referenced in Koran 4:82:

"Do they not then reflect on the Koran? Had it been from anyone other than Allah 'God', they would have certainly found in it many inconsistencies".

The bottom line is that new scientific discoveries would have proven it wrong in a major way had it been authored by an illiterate man riding on donkeys in a primitive 'premodernity time frame' society in desert Arabia in the sixth century.

This book even has answers for those subconscious thoughts in the advanced nations. The immortal intelligence is illustrating this point 'conservatively': if this book is from God indeed, its followers would not have been dependent on 'advanced' nations for technology. They would possess the 'know-how' to manufacture their own Boeing and Airbuses and the like. The truth is, they need to import everything—from the mobile phones they communicate with to the medicines they rely on for their health and general survival. Like I said, I would like to keep this at minimalist deliberation, so I will leave it at that. But the Koran interestingly explains how many people could be trapped in this subjective, egocentric line of thinking and how they wound-up losing their souls.

Whenever our statements and verses are proposed to them, they think to themselves, we are at a superior stature of civilization than them. Hence, we should have, if anything, 'the moral ground' to lead and be followed, referenced in Chapter Mary 19:73:

"When Our clear revelations are recited to them, the disbelievers ask the believers ˈmockinglyˈ, "Which of the two of us is better in status and superior in assembly?"

The Koran says, little do they realize this was a test for them, and we recorded what they said. In the end, every soul would come back to us 'alone', as we created them in the first place, referenced in 19:75-80:

"...And We will inherit what he boasts of, and he will come before Us all by himself."

The Koran further tells us in this first universe that not only every creature—'friend and foe'—will taste the mercy of God, but the happiness and superiority of people and nations will be alternated

305

among them, 'all created beings', referenced in Koran 3:140. If your nation is down now, others were inferior in the past. Allah alternates the ups and downs of people, nations and civilizations through the land so that Allah tests those whose belief is sincere and not wavering due to materialistic superiority:

"If you have received a wound, they have received a similar wound 'in the past. ' Such days We rotate among the people, so that Allah may know those who believe..."

The pro-Koranic people, i.e. Muslims, have been tested with materialistic and technological inferiority in the past when all the power and technology were in the hands of non-believers. All the while, they remained under persecution and mercy from those who took an antagonistic approach to early Muslims. This is known as the early Meccan period. Those who sided with the truth when it was not attractive, in other words, —those who could see through the fabric of the outward scare thread of ruins and thus converted to align themselves with the Singularity of Elohim/Allah during these least appealing times—were categorized as the highest level of believers towards Allah, known as 'Iman'. This is referenced in Koran 56:10-14: those who scored the highest 'the foremost people' majority of them would come from first-generation however a small number of them would be also emerging from the last generations. Koran 56:10-14:

"...The foremost and the foremost, They are the ones nearest, in the Gardens of Bliss. A multitude from earlier generations. And a few from later generations."

This special group of people, known as 'the Foremost', are those who can side with the truth beyond the short-lived, resistant commotion of glitter and glamour.

As Koran 3:140 illustrates, the days of superiority would alternate between nations and civilizations. Hence, pro-Koranic people, 'Muslims', would peak and lead global civilization in all scientific and technological spheres and, as such, contribute to human development up until that time, in what is known as the Golden Age of 'The House of Wisdom'—the first grand secular library in the world, where all ideas are exchanged and traded freely from the 7th century and blossomed until the 12th century.

During this period, inventions and developments of algorithms by Al-Khwarizmi, algebra and trigonometry by Al-Battani and Al-Farabi, the first standard medical textbook in the world, *The Canon of Medicine* (*Al-Qanun fi al-Tibb*) by Ibn Sina ('Avicenna'), and the first fully developed hospital in the world, the 'House of Healing' in Baghdad, would emerge from this civilization in the 8th century known as the Brmistant of Baghdad. The attire of graduation gowns and the term 'Alumni' are derived from 'uluum', which is the plural of knowledge in the Arabic language. The above are but few owed to this culture and civilization. Much the work referenced today from Greek philosophers like Plato, Aristotle, Hippocrates, Galen,

Euclid, and Ptolemy would have been lost had it not been for the preservation of this movement. All the while, the Western world was in what is known as the Dark Ages during this period.

Look, my dear reader, the point is that civilizations have emerged and died in the past, just as the verse is highlighting. Yes, hegemonic civilizations have changed hands many times over in history. Therefore, do not let your superiority of today, that is —if you are from advanced nations—blur your intellectual foresight, as civilizations have met their own demise in the past. So, too, will we, as individuals, one day meet our ends and thus return to Allah, as the Koran referenced above, the whole chronicle of rollercoaster was but a trial.

The Koran does not rely on dogma and faith to claim its legitimacy, unlike other religious books, nor does it depend on the indoctrination of people while they are minors and cannot differentiate right from wrong through parental and cultural authority. It also does not use the threshold of an artificial orbit as a preventive instrument, which declines to probe the authenticity of the religion in the form of blasphemy. This, in turn, emotionally blackmails truth seekers, preventing them from questioning and examining their belief systems. If you scrutinize the religious scripture you were told as a child is the word of an infallible God, you are labeled a heretic and/or blasphemous! Only those who are hiding something bar genuine investigative analysis. The Koran, on the other hand, not only welcomes investigation but provides scientific evidence for those who are sincere truth seekers like you and me.

As mentioned above, we will contemplate and answer the fundamental question through this part of our conversation. Can an illiterate man living in primitive Arabia in the sixth century guess these following accurate scientific statements? The claim that he may have copied from the Bible fails due to the fact that the Bible has demonstrated its own major scientific inconsistencies.

Expansion of the Universe

We know, or can readily access, one of the greatest scientific discoveries of this century include: the expansion of the Universe, first brought to light by the scientist Edwin Hubble in 1929. Further, Stephen Hawking, a self-proclaimed atheistic scientist, affirms this discovery as amounting to "a great intellectual revolution of the 20th century" in his book about the subject, *A Brief History of Time*. This phenomenon of rapid universal expansion was confirmed as recently as 2011 by supernova observer scientists Saul Perlmutter, Brian P. Schmidt, and Adam G. Riess, who were recognized for their discovery on the accelerating expansion of the universe. This led to their Nobel Prize and confirmed that, indeed, the universe is expanding at a much faster rate than previously known. Not only that but there is an unseen power called 'dark energy' which is causing this expansion, overriding any possible gravity slowdown that was anticipated in the past.

If the illiterate donkey rider in Arabia, Muhammad, took a risk and claimed that, yes indeed, the universe is expanding 1, 400 years ago—when no one could even remotely guess such a thing—and was found to be correct, would it not be sufficient evidence to ascertain his claim that he is that prophet, the 'Spirit of Truth' that Jesus prophesied would reveal to people new information? And when he says, "I hear this from the unseen power behind the universe, the Singularity Elohim/Allah, who is behind this expansion", would it not amount to exactly what science agrees is the unseen power conducting this rapid expansion, called 'dark energy'?

Digest what the verse says in the Koran, referenced in 51:47: **"We built the universe with ˹great˺ might, and We are certainly expanding ˹it˺".** In other words, the unseen Creator with His unseen power—'dark energy'—created the mighty, 'enormous' universe, and it is the same All-Powerful entity who is expanding it with His unseen powers. Now the question to you and me is: is he, not Muhammad 'peace be upon him' (a term devoted to prophets of God in the Koran), a true prophet due to the above fact? Regardless of what anyone says, there was no way a donkey rider in the premodernity era sixth-century desert Arabia, without a telescope, could make such a prediction and be found to be accurate today.

Embryology

It is not only interesting but also an important topic, as embryology is the actual certifiable process of our evolution—how we, as human beings, developed stage by stage until we became the final product we are at the moment of birth. This can prove to us that anyone claiming to have this information in the sixth century must have received it from God, since there was no resource available to detail such a chronology.

Stage One – The Koran, 23:13, states that human development begins when the male 'drop of fluid' makes contact with the female's 'reception egg', and a union is formed. Science calls this union a 'zygote', which is when the male sperm fertilizes the egg.

Stage Two – The Koran, 23:14, states the formation of an embryo appearing as a 'clinging clot' resembling a leech clinging, which modern science calls the terminology 'blastocyst', embedding onto the uterine wall as it attaches itself to the uterine wall. This is in precise concurrence with the Koranic account.

Stage Three – The Koran, 23:14, continues, stating that this is the stage where the bones are formed first, followed by the flesh, which encompasses the muscles, ligaments, and the like. Modern science calls this the terminology of 'fetus' and 'ossification, ' which is exactly consistent with the Koranic account.

Stage Four – The Koran, 23:14, describes this as the final stage, where the development of the baby is completed. This implies the final wiring of the brain and nervous system, the activation of the immune system, and the like, all synchronized as human development reaches its final stage. Modern science elaborates on this to mean the stage where all the nervous systems, the sensory organs, and other body organs are fine-tuned to work together as a coherent unit.

The Koranic description (23:12-14) states:

"And indeed, We created humankind from an extract of clay, then placed each ˹human˺ as a sperm-drop in a secure place. Then We developed the drop into a clinging clot, then developed the clot into a lump ˹of flesh˺, then developed the lump into bones, then clothed the bones with flesh, then We brought it into being as a new creation. So blessed is Allah, the Best of Creators."

Due to the bold and brave afore claim put forward by Muhammad, the illiterate man with his donkey in the 'premodernity' primitive society in Arabia, who lived between the fifth and sixth century AD and claimed God talked to him, presented a great opportunity to his critics to unmask his boastful fraud and 'catch his lies.' Therefore, many scholars in the embryology field scrutinized the above claim and chronology to show his fakery. These field experts included Professor Keith Moore, a founding member of the American Association of Clinical Anatomists (AACA) and a renowned professor in the division of anatomy in the Faculty of Surgery at the University of Toronto. He was in the greatest shock of his life—'an awakening alarm'—and he uttered upon examining the scientific facts of Muhammad (PBUH), "I was astonished by the accuracy of the statement recorded in the 7th century AD. "There is no alternative means with which to reject the claim that Muhammad must have reported from God, Allah. Dr Keith Moore affirmed the prophethood of Muhammad.

It was not only him; many further scholars in the field had similar jolting astonishments from the accuracy of Muhammad (PBUH). These included Professor Dr Marshall Johnson, a professor in Anatomy and Developmental Biology at the University of Philadelphia, who said [paraphrased meaning], No ordinary human being in the premodernity era of the sixth century could correctly predict the internal development of the human embryo taking place in a concealed environment to the naked human eye in pregnancy. "If I were to transpose myself into that era, knowing what I do today and describing things, I could not describe the things that were described". He further added, "I see no reason to refute that Muhammad must be developing this information from some other place—a higher being."

That is why the historian Thomas Holland, in his documentary *Islam: The Untold Story* (2012), was intrigued by how, out of nowhere, in the middle of an empty desert, a great civilization would rise, one that would supersede simultaneously both world empires at the time the Eastern Roman Empire and the Persian Sasanian Empire. As he visits the vast, empty desert of Arabia, Mr Holland contemplates in the documentary why God would choose His last prophet from 'here' where there is nothing—an empty, barren desert. No civilization, no literacy, no scholarly Biblical rabbis, no high priests, no Jerusalem Temple, no laboratory—nothing. He realizes that its emptiness, 'in and of itself' is its point of purity so that there is no alternative possibility to assume that the source of Muhammad could come from anything other than God, Elohim/Allah.

Remember, finding the truth does not automatically translate to its conformity 'acceptance', as many knowledgeable people may still reserve their willingness to submit to the truth, even when they know it. This is something we have seen in Lucifer (Iblis), the fallen angel who knew God intimately; the Rabbis and Pharisees of the Temple of Jerusalem against Jesus; and even Muhammad, whose some of his famous family members, namely his uncle and guardian Abi Talib, rejected following him, despite he would swear in his poem Muhammad 'my son' never lied, not even once in his lifetime and later the Koran reminding Muhammad that you do not guide whom you have an affection for at your mere wish alone, indicating that the 'test of life' is self-autonomous to each individual human being under their free will this is referenced in Koran 28:56:

"Indeed, [O Muhammad], you do not guide whom you love, but Allah guides whom He wills. And He is most knowing of the [rightly] guided."

The Beginning of the Universe –the Big Bang Theory

The fact that all existences in the universe go back to a single point, from a single outlet which rendered hot and high-density 'gas' to violently erupt—'bursting out'—and spring forth additional celestial existences, compounded by the fact that we can witness it, just like an event in a movie with a rewind button, was the highlight of our book discussed extensively above in Edwin Hubble's observatory evidence in 1929.

The allowance of this fact is no coincidence; it has been a passage from God for those who intend to ascertain the absolute reality of the Singularity 'Ahad/Echad' God Eloh/Allah, with no competing candidates, enshrined in all the first commandments of Abrahamic religions, irrespective of people enforcing it or not, with the temporary free will granted to expire at death. Since this is the Koran section, the chapter Ikhlas 'Sincerity' 112:1, Muhammad was commanded to **"inform your Lord 'the source' is the Singular."**

Further, for Muhammad, the illiterate donkey-riding shepherd in primitive, barren Arabia, to preemptively assert such a forecast 1400 years ago during the premodernity era with precise accuracy is another 'awakening alarm' to those who are sincerely interested in the factuality of our origin. Hence, the Koran asks a rhetorical question to give our hearts the 'electrical shock' it needs to reignite, just as the A&E hospitals perform when hearts deviate from their dispositional rhythmic straight path.

This is referenced in the Koran, 21:30: **"Do the skeptics not consider that the heavens and earth were ˹once˺ one mass, then We split them apart..."**

Life is based on water

At first glance, the above claim is not much of a shocker; almost any ordinary human living in the sixth century could estimate the causal link between life and water, which is why astronomers who are looking for signs of life in other clusters of planets focus on spotting water, hence making it synonymous with the phrase "where there is water, there is a sign of life". However, upon deeper analysis, this Koranic statement of 'every living thing is derived from water' unveils another point within it, just as the Singularity, because the two are connected in the same verse, as above, in Koran 21:30:

"Do the skeptics not consider that the heavens and earth were ⌜once⌝ one mass, then We split them apart? And We made every living thing from water, will they not believe?"

Then the question is: what does the Singularity have to do with water? What is this evidence of water being the common ingredient of every living thing revealing to us? The answer is that every living thing, in scientific terms, means organisms, which encompass the essential life form in every living body/entity. Therefore, this is applicable to a wide range of life forms, whether it is single-celled organisms such as bacteria to sophisticated beings such as multicellular organisms, including plants, animals, and, yes, humans as well.

Therefore, these cells are necessary mechanisms for the structural and functional components of life to exist and flourish. Science discovered that the water content of a cell amounts to between 70% and 90% in the 20th century by Ernst Overton and others. Without the advancement of the microscope, which led to the discovery of cells in the 17th century, cell existence, in general, would have been unknown to us, let alone an advanced understanding of its typologies, such as 'stem cells', which were learned in their human embryonic form as recently as 1998, and the like.

The Koranic intellectual attachment of water with the singular source of creation clearly points to the fact that we have a singular dominant ingredient in our bodies, 'all of us', from bacteria to plants, animals, and all the way to humans. We are from the Singularity, connected with the single point of the Big Bang's objective goal. In other words, water is yet another piece of evidence that we are all from not only the same source, but a Singular source, which takes the meaning of Allah for the impartial truth seekers.

Timescale of the Universe:Earth time (Human's time)versus Cosmic Time (God's time)

This topic is the one that my immortal intelligence finds to be as much nerve-wracking as it is inspiring to achieve the highest of possibilities due to the comprehension it unlocks once your attentive immortal intelligence grasps it.

The Koran informs us that God, who is on another timescale than that of human earthly time, has patience for His flagrant deniers who perpetuate their status quo in their predetermined lifespan given to them. They are emboldened even further in their antagonism due to that relative security, despite the fact that they are certain they will die thus loose the security. No names are needed to be leaked, but those who encompass our day and age include those atheist scientists and people of influential stature who blatantly claim there is absolutely no source behind creation—'there is no God'—as if they took a celestial holiday trip with their binoculars and peeked beyond the edge of the universe and found out, in a new discovery, that 'there is nothing there'. This, consequently, means this whole smart existence is, therefore, a game without any purpose behind it. They are as confident as a hallucinating eye interpreting facts, in their fancy words of misleading 'vacuous discourse' formulas such as, "The science community thinks rather 'guesses' such and such, " to confuse the layperson like you and me, as though their guessing of preferential opinion is an absolute scientific fact.

The Koran, henceforth alluding to the patience of God, asks why are they not ceased immediately, in other words, why their lifespan is not revoked instantaneously. It tells us in Koran 22:47: **"They challenge(us) to hasten the torment. And Allah will never fail in His promise. But a day with your Lord is indeed like a thousand years by your counting."**

This enlightens us that a day where the Lord is beyond the edge of the universe's timescale is tremendously different in that it is moving so slowly that a day amounts to a thousand years. This means, in our earthly minds, the timescale of where God is, amounts to approximately an hour being 42 years of our earthly time. Therefore, a minute would amount to approximately more than 8 months of our earthly time.

What is astonishing about cosmic time in comparison to Earth time is that it verifies the Koranic claim that 'time' infinitely slows down as one travels higher and higher in the universe. The closer one reaches to its highest points, near its high density or the powerful gravitational pull of black holes and the like, the slower time gets. This is a significantly slow-moving time, as Einstein theorized in his ground breaking theory of general relativity. For the average intellectual human being with immortal intelligence, there cannot be any contentious dispute on the fundamental ground of the mere factual timescale difference between how time moves on planet Earth in comparison to the furthest regions in our universe. Consequently, based on this premise, it unlocks the Koranic prioritization of the next life, which will be based on that stretched time clock. A winner in that life will enjoy his/her winnings 'spoils' infinitely larger than those of the Earth because the entire life on Earth would amount to between two to three hours, which is calculated to be between 80 to 120 years of earthly age in terms of lifespan.

That is the inspiration to be taken from the Koran and to prioritize your life accordingly. Also, take into account (as it goes without saying, may God forbid) that any losses, however minuscule on planet Earth, could also be infinitely stretched due to the timescale of the next life. The takeaway is that you and I should pray and work as smart as possible because we 'literally' cannot afford to be losers in such a timescale 'existence' where a day could be one thousand years of our current time on planet Earth.

The Sky's role of protective functionality

Yet another impossible piece of scientific knowledge from the sixth century, and Muhammad(PBUH), is the protective role of the sky for the planet Earth and its inhabitants. The Koran made it one of its evidences of validation. It is actually saying to us, look, this Koran, with this precise scientific depth, cannot possibly come from a human in the sixth century. Why are you not activating your immortal intelligence? Referenced in Koran 21:32: **"And We have made the sky a well-protected canopy; still, they turn away from its signs."** The fact that it is rendering advanced caution—"still people would turn away"—suggests that, in some cases, the grounds of rejection for the Koran are beyond the contest of whether it is true or not but are subjective, with disagreements about the purpose of life itself playing a needless role for the unfortunate rejecters.

For Muhammad, the illiterate orphan donkey rider in the desert of Arabia in the 6th century, to guess that the sky protects human inhabitants on planet Earth from the complex and sophisticated elements of space is hopelessly inconceivable. Yet, he draws our attention to it as evidence of his validity as a prophet of God. This is taking into account that human advancement would reach 13 centuries later when, in the 1930s, British physicist Sydney Chapman developed the theory explaining the inner workings and objective functions of the ozone layer. Along with other physicists such as Carl Friedrich Gauss and James Van Allen in respectively the 18th and 19th centuries, they discovered how Earth has an invisible shield known as the magnetic field, which offers ceiling 'canopy' protection to the planet Earth from harmful cosmic and solar rays and particles, just as Muhammad described 14 centuries ago.

When the Koran Corrected Science and the Bible

As the title suggests, there was an incident when the Koran and science, together with the Bible, clashed. However, it was science that 'gave way' and was proven wrong and corrected at a later stage, which also debunks any notion that Muhammad copied from the Bible.

Nearly three hundred years prior to the revelation of the Koran in the 2nd century AD, the Geocentric Model gained traction in the world of academia, as the mathematician and astronomer Claudius Ptolemy developed the model 'Almagest', which was extracted from the foundational views of Aristotle and Hipparchus. This model placed the Earth at the centre of the universe, stationery as the rest of celestial bodies, such as the Sun, Moon, planets, and stars are orbiting around it. Complying with this incorrect view in the name of science, church authorities supported this view since its inception from the 12th century by Thomas Aquinas all the way to the 16th century by Pope Paul V during the Renaissance era. This was based on Biblical statements in the Old Testament referenced in Psalm 104:5, which plainly expresses that the Earth is fixed and does not move: "He set the Earth on its foundations; it can never be moved."

The Koran, which is known to have been revealed during the lifetime of the Prophet Muhammad (PBUH), defied both the scientific community and the church fathers on the established doctrine of the Geocentric Model. This model was based on the fixed centrality of planet Earth in the universe, and its clear opposition is stated and referenced in Koran 21:33, which clearly omits itself to follow the idea of a fixed Earth proposed by the scientific community along with the church, and the Bible: **"And He is the One Who created the day and the night, the Sun and the Moon—each travelling in an orbit."** Later scientists, based on their own research, would align their views with the idea that the planet Earth is also in motion, as well as other celestial bodies. This would be known as the Heliocentric Model, conceived in the 16th century by Nicolaus Copernicus and later weighed in and supported by other scientists of the day, such as Galileo Galilei and Johannes Kepler, who conducted telescopic and mathematical research.

Therefore, it is clear that Muhammad was not copying from the Bible because if he had, he would have conformed to the well-established Geocentric model that the church had adopted. After all, the Koranic revelation occurred at the peak of the Geocentric model in the 6th century. However, the above is not to say the Koran has no similarities with the Bible, as the Koran itself claims it is the extension of the central theme of the Bible, which is the exclusive worship of Elohim/Allah, enshrined in the first commandment. With that said, the Koran claims that it corrects the Bible where errors were added by the scribes.

Iron 'not' an earthly material

For Muhammad(PBUH) not only to come up with but even to remotely contemplate that iron is not an earthly material but was brought to the planet Earth by meteorites that impacted Earth during its infancy period is nonsensical. In fact, it amounts to recklessly betraying of your immortal intelligence. The Koran draws our attention to this fact, which is proof within the proof, evidence within evidence. The Koran is informing us that just as iron is not an earthly material and was sent down, so too the Koran is not an earthly material—it was sent down. In other words, it is not manmade speech; neither Muhammad(PBUH) nor anyone else could be its author. Please digest this phenomenon yourself, referenced in Koran 57:25: **"Indeed, We sent Our messengers with clear proofs, and with them, We sent down the Scripture and the balance ˹of justice˺ so that people may administer justice. And We sent down iron with its great might, benefits for humanity, and means for Allah to prove who ˹is willing to˺ stand up for Him and His messengers without seeing Him. Surely Allah is All-Powerful, Almighty."**

It is interesting that the Koran asks us to ascertain God for ourselves, but not in the ordinary way of knowing objects and people. Rather, it is in a way that only befits His majesty. We were not asked to know the Creator of the universe as we know each other or other beings for whom we conceive an image, as well as put dimensions and parameters around them. Therefore, we have some precise image of how tall, wide, etcetera our object is, including whether they are slim or wide-built. Our intellectual challenge before us has two underpinnings: (a) that we are grateful to Him, and (b) that we do not sell out our intelligence to anything lower than the capability of the creation of the 'universe' heavens and the earth. Henceforth, we must dismiss any proposal other than such a capability, such as humans and/or anything like or less than humans.

The Double Entry Preservation of the Koran as a Wondrous Evidence

How any historical book is preserved is not only vital for the accuracy of the message it aims to transmit, but also it depicts to its audience and the general public its legitimacy, and how much it can be trusted. So, it must be a truthful document. The scriptural religious books must show an even higher degree of authenticity and reputability in their compilations and scribes than academic books due to the magnitude of what they are arguing for — which is our complete surrender and obedience. Why would any critically thinking person surrender to dogma, faith and blind following?

Consequently, for the immortal intelligence, there is nothing more painful than upholding and obeying the 'dos and don'ts' of religious doctrine for so long, only to discover that it was, after all, man's figment of imagination for whatever reason. Although you can bet with all your money, it would be either or both, 'wealth and power grab' that made these scribes to alter the scripture. So much ritual labor to give, so many sacrifices to make, so many charities to give, and then to find out it was all someone pranking you — except in this prank, you were the fool whose money and servitude were taken for a ride. This is a grievously agonizing experience for any human being to go through. This is further referenced in the Koran 18:103-104: **"Say, ˹O Prophet,˺ "Shall we inform you of who will be the biggest losers in deeds? ˹They are˺ those whose efforts are in vain in this worldly life, while they think they are doing good."**

In order to avoid becoming a victim of the above scenario, the immortal intelligence will scrutinize the compilation of the Koran in comparison with its counterparts, such as the books of the Bible.

The Koran is just a different breed altogether when it comes to the sensitivity of preservation. It lays the accusation that other religious books God had revealed were altered, 'tampered with', over time, referenced in 2:79: **"So woe to those who distort the Scripture with their own hands, then say, 'This is from Allah' — seeking a fleeting gain! So woe to them for what their hands have written, and woe to them for what they have earned."** But that is not enough. It asks the reader, would you be a fool enough to think God would not call out such alterations?

In other words, did you expect God — being God — would be silent all the while people are changing His message of unconditional surrender to the Singularity with no intermediary? Additionally, the Koran says this new revelation, i.e., the Koran will be out of the hands of scholars of the religion to alter it any longer, referenced in 15:9: **"Indeed, it is We who sent down the Koran, and indeed, We will be its guardian."** This verse is a unique claim that this book, being the last scripture for all the nations in the world as we were headed into the globalization of the world, is a self-preserving book — 'God Himself assumes its guardianship' — as such, would be unlike other books which preceded it in the past.

Such a bold assertion, is it not? Hence, this book would not only be in written form but also memorized as well, so if the writers made even unintentional mistakes, the reciters would catch the mistake and correct it 'vice versa'. Thus, it works much like the 'double entry' accounting system — 'guardianship' Assets = Liabilities + Equity, which is purported to preserve the balance and accuracy of the financial activity of the company. As the name suggests, each transaction must be recorded twice; hence, two entries would be recorded for each transaction. Therefore, inevitably, the two entries must match; otherwise, there is an error. The Koranic 'double-entry' process preceded that of the accounting model by at least 8 centuries before Luca Pacioli explained such a principle in the 14th century.

As for the Hebrew Bible, The Pentateuch, it is unanimously agreed that it was edited and prepared by unknown people as we know it today. The Torah, comprising Genesis, Exodus, Leviticus, Numbers, and Deuteronomy, was written at least five to twelve centuries after Moses (PBUH), its central figure. Further, it is commonly agreed that Moses did not write everything mentioned in these books, as cross-referenced in Bible scholars such as Joel B. Baden in his research on the subject, *The Composition of the Pentateuch: Renewing the Documentary*. Others, such as John Van Seters, Richard Elliott Friedman, and Bart Ehrman, have also added their voices to this view.

Additionally, as for the New Testament, churches such as the Catholics and Protestants do not even agree on the number of books that make up the authoritative Bible. Is it 66 or 73 books?

The earliest of them was written at least 60 to 90 years after Jesus Christ, its central figure. However, the final editing and official making of the Bible took place over three hundred years after Jesus at the Council of Nicaea in 325 AD. In the 16th century, at The Westminster Confession, 13 Bible books, otherwise known as Apocryphal Bible, were simply deselected from the Bible at will.

As the public was unaware, but those in power knew, Donald Rumsfeld, George Bush, and Tony Blair knew there were no weapons of mass destruction (WMD) in Iraq in 2003. Similarly, the Bible, while it is widely known that these central figures such as Moses and Jesus did not directly write these Books yet, the public is unaware that fact, as Rumsfeld would call it 'the unknown of the known' as his justification of blatantly lying to the public of starting the Iraq war.

Additionally, neither the Bible was endorsed by these central figures, nor was it written under their supervision. So much so that no one can verify who the Bible writers were. Matthew, for example, has no history, no known last name, and no family, as well as Mark, Luke, and John. St Paul is the only exception of the writers, who had to change his name from 'Soul' in order to cleanse 'whitewash' his trace of hatred towards Christianity and his documented persecution of early followers of Jesus before he self-proclaimed his 'apostolate'—something we discussed in depth in our earlier conversation.

Prophet Muhammad (PBUH) applied the double-entry system as he was receiving the Koran. He allocated his scribes to write it down while concurrently, others were memorizing its melodious recital. These allowed the double authentication 'entry' to take effect in practice under the supervision of Muhammad (PBUH) himself. Furthermore, all his community would pray with him five times a day, being exposed to the recitation of the Koran directly from him, thus enjoying 'mass-witnessing' by thousands of people. Additionally, he had a group of dedicated memorizers and students, many of whom lived in his mosque, known as 'Ahlu Suffa', meaning people who lived under the shade of the mosque in the hot weather of Arabia. They were 'poor students' who had unhindered access to the Prophet.

As a result of the clever strategy of the double-entry system, at the death of the Prophet, the Koran was spread around in society both in written form 'parchment manuscripts' and recital form. Within a little more than a year after his passing, an incident occurred in which dozens of Koran memorizers (Hufaaz) lost their lives, triggering the need for the first Caliph, Abu Bakr, to compile the written form into one document.

This task was commissioned to one of the Prophet's young dedicating scribes, Zayd ibn Thabit, with the strict requirement that each verse must be confirmed by two independent writers who would present their record from their archive and two memorizers (hafiz) who would give under oath testimony that they had heard the same verse, word for word, from the Prophet and memorized it. Remember, all the while, Mr Thabit was both a scribe and a memorizer himself. As Mr Thabit would report, there is only one verse in the entire Koran which failed to acquire all four witnesses' requirements, attaining three witnesses in lieu, which he passed it to include in the Koran compilation, as he became a witness to it in compliance with the existing previous Prophet Muhammad's character witness statement for the third person who brought the verse from his memory in the Hadith authority, referenced in authentic Bukhari 4986.

There cannot be a debate, to the impartial eye, that the preservation of the Koran, in comparison to that of the Bible, is a universe apart. With that said, this and the above evidence forwarded above should not be translated to mean verbatim that the Bible is an entirely bogus document. Quite the contrary, because the Koran itself gives the endorsement that the original Torah given to Moses (PBUH) and Injil ('Gospel') given to Jesus are authentic scriptures from the Father, God Elohim/Allah, as stipulated in the first commandment. Therefore, what the Koran and other scholars are saying is that, over time, the word of God was contaminated with the word of man. Whether intentional or not is not our primary concern and would not affect the fact that these books have been diluted and transformed from the intended meaning the father had revealed for in the first place; therefore, what matters is that contamination had taken place. Consequently, the biblical books cannot be taken as the verbatim 'inerrant' word of God, if you were not already informed about the fact.

As for the Koran, the immortal intelligence will conclude with a powerful, self-evident statement, which is the last recorded verse that Prophet Muhammad (PBUH) ever recited in public before he passed away, which asks you and me a question, referenced in Koran 77:50: **"So what message after this ˹Koran˺ would they believe in?"**

The Beginning

The immortal intelligence rests its intellectual journey in this section, which we will call "the beginning", where others would title it "the conclusion", The beginning symbolizes multiple themes of hope and inspiration in the nature of moving forward: today is another beginning, another dawn, another day, another hour, an opportunity to redeem yourself, produce your best yet, and become the best version of you again. Even as you slip and suffer setbacks, refuse to allow your faults to define who you are, as the Koran regards as 'the foremost' class of people, and further carry that audacity of good expectations and positive thought of your Lord under all circumstances. These words were the valediction of Prophet Muhammad (PBUH) in his passing moments.

The aim of the above analytical engagement was not to debunk any culture of people or race but to question the facts, as complex and convoluted as they are packaged for us, from religious doctrines to 'rocket science' formulas. We have learned, yes, our so-called average intellect is capable of decoding and seeing the truth through the multifarious layers of misconceptions and unsubstantiated theories. In fact, my immortal intelligence became transcendent to race and cultural preferences—the things which divide us as global citizens and spark hatred in the world.

My immortal intelligence literally sees itself, as Prophet Muhammad (PBUH) said in his farewell speech, as a reflection of my fellow human beings, as we are 'Adaminians' , which means we all go back to the same ancestor, 'Adam', and would strive to give my fellow human beings the same rights as I would like to receive for myself and my loved ones, as Jesus (PBUH) taught us through this Book.

Muhammad and Jesus 'peace be upon them' (PBUT) are two men who fulfilled each other despite the baseless divisions we hear all the time. Furthermore, through the reflective research of this book, it became intrinsically lucid to my immortal intelligence that the discourse suggesting religions are the villains of our global sufferings and conflicts is not simply true. They are quite the opposite.

The scriptures we have had the opportunity to evaluate all embodied meaning well and were intended to be considerate towards your fellow human beings. But rather, it is those who interpret them with ill intentions in the extreme who are the cause of divisions and sufferings amongst nations and neighbors. Unfortunately, they take advantage of our ignorance of the scriptures, regardless of how superbly trained we may be in our skilled jobs.

Above all, what this research proved to you and me is something we were already innately aware of but had not brainstormed quite like the above. The fact is that we are the same species of humans, created predominantly from water, as we have seen above. Therefore, we all go back to the same source, who we found out to be Allah/Aloe/Ella/Elohim. Whatever your language, whatever your dialect, whatever vernacular or scientific justification you substantially consider, it inevitably leads to the fact that all roads converge onto the same unseen, All-powerful source behind the universe—the supreme 'Creator.' There is no other logical, workable, or provable alternative answer other than Allah, the Singularity. Go ahead and try other proposals: Jesus! The one who tells us, "I am a prophet, son of man, created 'being'" (please refer above in our discussion); Buddha! The one who tells us, "It is not my business to answer the question" or a 'Random Chance!' A fancy way of claiming 'nothing'. If that is the case, then nothing should be in existence including you and me. Since when would our immortal intelligence be bamboozled, 'duped', into accepting that nothing can produce something?

Now, what you do with the above factual information, which we came to through deductive reasoning, is up to you. The awakening of your immortal intelligence contained in this book was never to sway you in any direction, oh no, my dear friend, it has been nothing but your call for your soul! It was only to provide you with the data and the facts beyond the impediments of fog and divergence the artificial orbit lays before you and me so that we do not see the baseline. Furthermore, it aims to simplify manmade, unnatural doctrine(s) and models, so the road to paradise—'success'—is clear in front of you and me, should we wish to embark on it with our free will.

This is the last code I would like to leave you with: "My dear friend, do not surrender your immortal intelligence at any cost. Realize it has always been the 'Singularity/Ahad/Echad' Allah all along."

Salaam / So-long / Shalom.

www.ingramcontent.com/pod-product-compliance
Lightning Source LLC
Chambersburg PA
CBHW060758120626
46557CB00001B/22